Approximate Solution of Random Equations

Editor
A. T. BHARUCHA-REID
Department of Mathematics
Wayne State University

NORTH HOLLAND
New York · Oxford

Elsevier North Holland, Inc.
52 Vanderbilt Avenue, New York, N.Y. 10017

Distributors outside the United States and Canada:

Thomond Books
(A Division of Elsevier/North-Holland Scientific Publishers, Ltd.)
P.O. Box 85
Limerick, Ireland

Library of Congress Cataloging in Publication Data

Special Session on Approximate Solution of Random Equations, Atlanta, 1978.
Approximate solution of random equations.

(North Holland series in probability and applied mathematics)

"Many of these papers were presented at the 752nd Meeting of the American Mathematical Society, Atlanta, Georgia (3–8 January 1978) as part of a Special Session on Approximate Solution of Random Equations."
Includes bibliographical references and indexes.
1. Stochastic differential equations–Addresses, essays, lectures. 2. Stochastic integral equations–Addresses, essays, lectures. 3. Approximation theory–Addresses, essays, lectures. I. Bharucha-Reid, Albert T. II. American Mathematical Society. III. Title. IV. Series.
QA274.23.S65 1978 519.2 79-16107
ISBN 0-444-00344-4

Manufactured in the United States of America.

Approximate Solution of Random Equations

NORTH HOLLAND SERIES IN

Probability and Applied Mathematics

A. T. Bharucha-Reid, *Editor*
Wayne State University

AN INTRODUCTION TO STOCHASTIC PROCESSES
D. Kannan

STOCHASTIC METHODS IN QUANTUM MECHANICS
S. P. Gudder

APPROXIMATE SOLUTION OF RANDOM EQUATIONS
A. T. Bharucha-Reid (ed.)

Contents

Preface vii

List of Contributors ix

GEORGES A. BÉCUS
Successive Approximations Solution of a Class of Random Equations 1

W. E. BOYCE
Applications of the Liouville Equation 13

P. S. CHANDRASEKHARAN and A. T. BHARUCHA-REID
Approximate Solution of Fixed Point Equations for Random Operators 27

P.-L. CHOW
Approximate Solution of Random Evolution Equations 37

STUART GEMAN
A Method of Averaging for Random Differential Equations with Applications to Stability and Stochastic Approximations 49

R. KANNAN
Stochastic Approximation and Nonlinear Operator Equations 87

HAROLD J. KUSHNER
Approximation Methods for the Minimum Average Cost per Unit Time Problem with a Diffusion Model 107

MELVIN D. LAX
Obtaining Approximate Solutions of Random Differential Equations by Means of the Method of Moments 127

M. Z. NASHED and H. W. ENGL
Random Generalized Inverses and Approximate Solutions of Random Operator Equations 149

W. J. PADGETT
Approximate Solution of Random Nonlinear Equations by Random Contractors 211

TZE-CHIEN SUN
A Finite Element Method for Random Differential Equations 223

Author Index 239

Subject Index 241

Preface

At present, the theory of random equations is a very active area of mathematical research, and applications of the theory are of fundamental importance in the formulation and analysis of various classes of operator equation that arise in the physical, biological, social, engineering, and technological sciences.

The numerous theoretical methods that have been developed for solving deterministic operator equations have, in the main, been concerned with establishing the existence and uniqueness of solutions. Methods for solving random equations must establish not only existence and uniqueness, but also the measurability of the solutions. This is one of the essential differences between methods for solving deterministic equations and random equations. Another important difference is that the solution (exact or approximate) should be in a form that permits the study of its statistical properties.

In recent years an increasing number of papers and books have appeared that are devoted to the development of methods for obtaining approximate solutions of deterministic linear and nonlinear operator equations. In particular, the literature on approximate solution of differential and integral equations is very extensive. Since the 1950s, when the first systematic studies on random equations were initiated, it has been clear that approximation methods would play an important role in the study of the solutions of random equations and their statistical properties. Indeed, the fundamental papers of Itô, and Špaček and Hanš utilized fixed point (in particular, successive approximation) methods to obtain solutions of random differential equations and probabilistic analogues of classical integral equations.

During the past 10 years research workers in the field of random equations have been devoting a great deal of attention to the development of methods for obtaining approximate solutions of various classes of random equation. This area of research, which utilizes current results from numerical functional analysis and the theory of stochastic processes, and employs computers for numerical studies, is a fascinating one. The results obtained are of theoretical interest as well as of importance in the applied mathematical sciences.

The purpose of this book is to present a survey of recent work concerned with approximation methods for solving random equations. The 11 papers contained in this volume give a good account of the types of problem being studied, and the methods used, by research workers in this area. Many of these papers were presented at the 752nd Meeting of the American Mathematical Society, Atlanta, Georgia (3–8 January 1978) as part of a Special Session on Approximate Solution of Random Equations. In order to expedite the publication of these papers, this volume in the North Holland Series in Probability and Applied Mathematics was prepared from camera-ready manuscripts. We hope that this book will stimulate research in this important branch of probabilistic analysis, and that in the near future monographs and books will be published that present systematic accounts of various approximation and computational methods for solving random equations.

A. T. Bharucha-Reid
Series Editor

List of Contributors

GEORGES A. BÉCUS
University of Cincinnati

A. T. BHARUCHA-REID
Wayne State University

W. E. BOYCE
Rensselaer Polytechnic Institute

P. S. CHANDRASEKHARAN
Wayne State University

P.-L. CHOW
Wayne State University

H. W. ENGL
Johannes-Kepler-Universität, Linz, Austria, and University of Delaware

STUART GEMAN
Brown University

R. KANNAN
University of Texas, Arlington

HAROLD J. KUSHNER
Brown University

MELVIN D. LAX
California State University, Long Beach

M. Z. NASHED
University of Delaware

W. J. PADGETT
University of South Carolina, Columbia

TZE-CHIEN SUN
Wayne State University

Approximate Solution of Random Equations

Successive Approximations Solution of a Class of Random Equations

Georges A. Bécus

Department of Engineering Science
University of Cincinnati, Cincinnati, Ohio 45221

1. Introduction: Motivation
2. Successive Approximations Solution
3. Properties of the Successive Approximations Solution
4. Particular Case: The L_2 Case
5. Problems (i), (ii), (iii) Revisited

References

1. INTRODUCTION: MOTIVATION

Consider the following problems.

(i) Steady-state diffusion in an isotropic medium where the diffusive properties are space and concentration dependent. Assume the space and concentration dependences of the diffusion coefficient are separable; the governing equation for such a process is then

$$\text{div}[K(C)\text{grad}C] = G, \text{ in } D$$

where

$$K(C) = k(x)k_1(C),$$

which upon using the transformation of the dependent variable

$$u(x) = [k(x)]^{1/2} \int_{C_o}^{C} k_1(\alpha)d\alpha,$$

takes the form (Cf. [1], Section 2)

$$\nabla^2 u - q(x)\, u = g, \text{ in } D. \tag{1.1}$$

The partial differential equation (1.1) can further be transformed into the Fredholm integral equation of the second kind

$$u(x) = u_o(x) - \int_D \Gamma(x,\xi)q(\xi)u(\xi)dv(\xi) \tag{1.2}$$

where $\Gamma(x,\xi)$ is the Green function of D and u_o is the solution to $\nabla^2 u_o = g$ in D satisfying the auxiliary (boundary) conditions which must be appended to equation (1.1).

(ii) Heat conduction in a medium occupying a region V in which heat is being produced (or absorbed) at a rate proportional to the local temperature (e.g. heat conduction in a thin rod with radiation from the lateral surface). Here the governing equation is

$$\frac{\partial u}{\partial t} - \nabla^2 u + q(x,t)u = g(x,t),\ (x,t)\varepsilon\ Vx(0,T) \quad . \tag{1.3}$$

It can be written as the integral equation

$$u = u_o - \int_0^t \int_V \Gamma(x,t;\xi,\tau)q(\xi,\tau)u(\xi,\tau)dv(\xi)d\tau\ , \tag{1.4}$$

where Γ is the causal Green function for the heat equation in V and u_o is the solution of $(\partial/\partial t-\nabla^2)u_o = g$ in $V\times(0,T)$ satisfying the auxiliary (boundary and initial) conditions which must be appended to equation (1.3).

(iii) Determination of the transverse displacement w at the hinge of a system of two identical hinged bars, under an axial load η and whose tranversal motion is prevented by a linear viscoelastic spring connected to the system at the hinge. It can be shown [2] that the governing equation is the integral equation

$$[1-\eta(t)]w(t) = g(t) + \int_0^t f(t,\tau)\eta(\tau)w(\tau)d\tau , \tag{1.5}$$

where f is the creep function of the spring and g accounts for transversal loads as well as initial displacements. With $u=(1-\eta)w$, $u_o=g$ and $q=\eta(1-\eta)^{-1}$ equation (1.5) becomes the Volterra integral equation

$$u(t) = u_o(t) + \int_0^t f(t,\tau)q(\tau)u(\tau)d\tau . \tag{1.6}$$

Each of the above three problems has thus been reduced to the study of an equation [(1.2), (1.4), and (1.6) respectively] of the form

$$u = u_o + T[qu] . \tag{1.7}$$

Many other problems can be cast into the form (1.7). In some cases, unpredictable fluctuations in the properties of and/or inputs to the system under consideration will warrant consideration of a random version of equation (1.7), viz.

$$u = u_o(\omega) + T[q(\omega)u], \text{ a.s.} \tag{1.8}$$

where $\omega\varepsilon(\Omega,A,P)$, some suitable probability space. The remainder of this article is devoted to the study of equation (1.8).

In Section 2 we define a weak solution to equation (1.8). Its existence, uniqueness and construction by a method of successive approximations under sufficient conditions on T and q are obtained. The properties of this solution are investigated in Section 3. In Section 4 the results of Sections 2 and 3 are specialized for an important particular case: the L_2 case. Finally, in Section 5, some results concerning the solution of the three problems of this section will be presented. Before proceeding with this plan however, some remarks are in order.

Remark 1.1. In examples (i)-(iii) the operator T is an integral operator. This needs not be the case for the results of Sections 2 and 3 to hold as long as T satisfies the requirements imposed there. However, the approach followed, in particular Definition 2.1, is motivated to a large extent by the case where T is an integral operator (Cf. Remark 2.2 below).

Remark 1.2. Physical situations leading to and/or justifications for considering (1.8) rather than (1.7) for problems (i) and (ii) can be found in [1,3] and [4] respectively. For problem (iii), both the deterministic equation (1.7) and its random counterpart (1.8) were consider in [2] (Cf. also [5], Section 4.4D); equation (1.8) comes about when the axial load η as well as the transversal load and/or initial displacements are random.

2. SUCCESSIVE APPROXIMATIONS SOLUTION

In order to give a precise formulation to equation (1.8), we introduce the following. Let $L_p(\Omega) = L_p(\Omega,A,P)$, $1 \leq p \leq \infty$, denote the usual Lebesgue spaces of (classes of) real measurable functions on Ω with the norm $||f||_p = \{\int_\Omega |f|^p dP\}^{1/p}$, $f \in L_p(\Omega)$, $1 \leq p < \infty$ and $||f||_\infty = \text{ess sup } |f|$, $f \in L_\infty(\Omega)$.

Let $D \subset R^n$ and let $B(D)$ denote any Banach space of functions f: $D \to R$ equipped with a norm $|| \cdot ||_D$ satisfying $||f||_D \leq ||g||_D$ if $f \leq g$ pointwise in D. We will consider mappings g: $D \to L_p(\Omega)$ such that $||g||_p \in B(D)$. We denote $B(D;L_p(\Omega))$ the space of all such mappings; it is easily verified that $|| \cdot ||_{pD} = || \, || \cdot ||_p ||_D$ defines a norm on $B(D;L_p(\Omega))$.

Definition 2.1. Let T: $B(D) \to B(D)$ be a continuous linear operator. Let $u_o \in B(D;L_p(\Omega))$ and let $q \in B(D;L_\infty(\Omega))$. A function u: $D \to L_p(\Omega)$ is said to be a weak (or generalized) solution of equation (1.8) in D if

(i) $u \in B(D;L_p(\Omega))$

(ii) $u^*(u) = u^*(u_o) = T[u^*(qu)]$ (2.1)

for each continuous linear functional u^*: $L_p(\Omega) \to R$.

Remark 2.2. As mentioned earlier (Remark 1.1), Definition 2.1 is motivated by the case when T is an integral operator. In that case T(qu) is the integral of a random function. Following (and generalizing) Karhunen's definition of the integral of a random function [9], we define T(qu) as the unique element $w \in L_p(\Omega)$ pointwise in D such that for all $u^*, u^*(w) = T[u^*(qu)]$. Obviously this interpretation imposes some restrictions on T. For each particular T one has to verify that such a w does indeed exist.

Remark 2.3. Definition 2.1 makes sense since $q(x)\, u(x) \in L_p(\Omega)$, $\forall\, x \in D$ and so $u^*(qu) \in B(D)$. Indeed, in view of Holder's inequality $||u^*(qu)||_D \leq ||v||_{p'} M_q ||u||_{pD} < \infty$ where v is the unique element of $L_{p'}(\Omega)$, $p^{-1} + p'^{-1} = 1$, which can be identified with u^*, and $M_q = \sup_D ||q||_\infty$.

Remark 2.4. Equation (2.1) is a deterministic equation whose solution $u^*(u)$ is going to be constructed below by a method of successive approximations. The determination of $u^*(u)$ will not in general allow for that of u; its expectation and, if $p \geq 2$, its covariance will, however, easily be obtained from $u^*(u)$ by taking for u^* the linear functional which can be identified with $1 \in L_{p'}(\Omega)$ and $u \in L_{p'}(\Omega)$ (if $p \geq 2$) respectively. From now on we will assume that $p \geq 2$ whenever necessary.

Let us define a sequence $\{u_m\}_{m=0}^{\infty}$ by

$$u^*(u_m) = u^*(u_0) + T[u^*(qu_{m-1})], \quad m=1,2,... \tag{2.2}$$

It is clear that $\{u_m\} \subset B(D;L_p(\Omega))$. Indeed if $u_{m-1} \in B(D;L_p(\Omega))$, $u^*(u_m)$ as given by (2.2) is in $B(D)$. Furthermore, the right hand side of (2.2) defines a bounded linear functional on $L_{p'}(\Omega)$ pointwise in D. The Riesz representation theorem guarantees then the existence of a unique $u_m \in L_p(\Omega)$ pointwise in D satisfying (2.2). Since $u_o \in B(D;L_p(\Omega))$ by assumption, the induction argument shows that $u_m \in B(D;L_p(\Omega))$ for every m.

Now if $v \in L_{p'}(\Omega)$ is identified with u^* then $u^*(qu_{m-1}) = v^*(u_{m-1})$ where v^* is identified with $q\,v \in L_{p'}(\Omega)$ and so

$$u^*(q\,u_{m-1}) = u^*(qu_o) + T[u^*(q^2u_{m-2})]$$

which can be substituted in (2.2), and so on, to yield

$$u^*(u_m) = \sum_{j=0}^{m} T^j[u^*(q^ju_o)] , \tag{2.3}$$

where $T^o = I$, the identity operator on $B(D)$, and $T^j(\cdot) = T[T^{j-1}(\cdot)]$, $j=1,2,...$

We now show that the series (2.3) converges in $B(D)$. In view of the Cauchy convergence criterion it suffices to show that $||u^*(u_s) - u^*(u_r)||_D \to 0$ as $r, s \to \infty$. We have for $r < s$

$$||u^*(u_s) - u^*(u_r)||_D = ||\sum_{j=r+1}^{s} T^j[u^*(q^ju_o)]||_D$$

$$\leq \sum_{j=r+1}^{s} ||T^j[u^*(q^ju_o)]||_D$$

$$\leq \sum_{j=0}^{\infty} ||T^j[u^*(q^ju_o)]||_D$$

$$\leq \sum_{j=0}^{\infty} |||T|||^j\,||q||^j\,||v||_{p'}\,||u_o||_{pD} \tag{2.4}$$

where $|||T|||$ is the norm of T, and $||q||$ and v are defined below. The last step in (2.4) follows from

$$||T^j[u^*(q^j u_o)]||_D \le |||T|||^j ||u^*(q^j u_o)||_D ,$$

$$||u^*(q^j u_o)||_D \le ||v||_{p'} ||q^j u_o||_{pD}$$

for some $v \in L_{p'}$, identified with u^*, and

$$||q^j u_o||_{pD} \le ||q||^j ||u_o||_{pD}$$

where

$$||q|| = \sup_{f \neq 0} ||qf||_{pD} / ||f||_{pD} .$$

Inequality (2.4) shows that a sufficient condition for the convergence of (2.3) in $B(D)$ is

$$|||T||| \; ||q|| < 1 . \tag{2.5}$$

Unfortunately, convergence in $B(D)$ does not in general imply convergence in $B(D;L_p(\Omega))$. If however for a given T, the series $\sum_{j=0}^{\infty} T^j[u^*(q^j u_o)]$ can be shown to define a bounded linear functional on $L_{p'}(\Omega)$, (cf. Section 4 below for an important class of T for which this is true), we can invoke the Riesz representation theorem to conclude the existence of a unique $u \in L_p(\Omega)$ pointwise in D such that for all u^*

$$\lim_{m\to\infty} u^*(u_m) = u^*(u) = \sum_{j=0}^{\infty} T^j[u^*(q^j u_o)] \tag{2.6}$$

in $B(D)$. Since $u \in L_p(\Omega)$ pointwise in D and $u^*(u) \in B(D)$ for each u^*, $u \in B(D;L_p(\Omega))$.

Since in view of the linearity and boundedness of T, $T[u^*(qu_m)] \to T[u^*(qu)]$ in $B(D)$ we have proved in this case the following:

Theorem 2.5. Condition (2.5) is sufficient for the existence of a unique solution to equation (1.8) in the sense of Definition 2.1.

Remark 2.6. Theorem 2.5 gives a sufficient condition under which the deterministic Neumann series (2.6) converges. While trying to obtain a necessary and sufficient condition is hopeless in general, it may be possible in some cases to prove convergence, even when (2.5) does not hold, in particular for the series defining expectation and correlation (see Section 3 below).

Remark 2.7. Condition (2.5) can be replaced by the stronger condition

$$|||T|||\ M_q < 1 \tag{2.7}$$

since $||q|| \le M_q$. (2.7) will in general be easier to verify than (2.5) for given T and q.

3. PROPERTIES OF THE SUCCESSIVE APPROXIMATIONS SOLUTION

(i) The error resulting from truncating series (2.6) at the m-th term satisfies the following estimates

$$||u^*(u_m) - u^*(u)||_D \le ||v||_{p'} ||u_o||_{pD} \frac{||q||^{m+1}|||T|||^{m+1}}{1 - ||q||\ |||T|||} \tag{3.1}$$

$$\le ||v||_{p'}\ ||u_o||_{pD} \frac{M_q^{\ m+1}\ |||T|||^{m+1}}{1 - M_q\ |||T|||}\ . \tag{3.2}$$

Indeed, the estimate in (3.1) is an estimate for the remainder of the geometric series (2.4) while (3.2) follows from (3.1) and $||q|| \le M_q$.

(ii) The expectation of the solution is given by the series

$$E\{u\} = \sum_{j=0}^{\infty} T^j [E\{q^j u_o\}] \tag{3.3}$$

with the following error estimates

$$||E\{u_m\} - E\{u\}||_D \le ||u_o||_{pD} \frac{||q||^{m+1}\ |||T|||^{m+1}}{1 - ||q||\ |||T|||} \tag{3.4}$$

$$\le ||u_o||_{pD} \frac{M_q^{\ m+1}\ |||T|||^{m+1}}{1 - M_q\ |||T|||}\ . \tag{3.5}$$

Equations (3.3-5) follow from (2.6, 3.1-2) respectively and Remark 2.4.

(iii) The correlation function of the solution is given by the double series

$$R_u = \sum_{j=0}^{\infty} T^j\ [\sum_{k=0}^{\infty} T^k\ (R_{uo}^{jk})]\ , \tag{3.6}$$

where $R_{uo}^{jk}(x_1,x_2) = E\{q^j(x_1)u_o(x_1)q^k(x_2)u_o(x_2)\}$ and which follows from (2.6) and Remark 2.4.

4. PARTICULAR CASE: THE L_2 CASE

In this section we look at the important particular case when T is an integral operator and $B(D) = L_2(D)$. Letting $K(x,\xi)$ denote the kernel of the operator T, we notice that in this case, if K is a Hilbert-Schmidt (or L_2) kernel, a crude estimate of $|||T|||$ is provided by

$$|||T|||^2 \leq \int_D \int_D |K(x,\xi)|^2 \, dx d\xi \, . \tag{4.1}$$

If furthermore there exists a finite M_k such that

$$\int_D |K(x,\xi)|^2 \, dx < M_k^2 \tag{4.2}$$

then it can be shown that the method of successive approximations of Section 2 converges uniformly in D. In this case the error estimates (3.1-2) and (3.4-5) become

$$||u^*(u_m) - u^*(u)||_D \leq ||v||_{p'} \, ||u_o||_{pD} \, M_k \, \frac{||q||^{m+1} \, |||T|||^m}{1 - ||q|| \, |||T|||} \tag{4.3}$$

$$\leq ||v||_{p'} \, ||u_o||_{pD} \, M_k \, \frac{M_q^{m+1} \, |||T|||^m}{1 - M_q \, |||T|||} \, , \tag{4.4}$$

and

$$||E\{u_m\} - E\{u\}||_D \leq ||u_o||_{pD} \, M_k \, \frac{||q||^{m+1} \, |||T|||^m}{1 - ||q|| \, |||T|||} \tag{4.5}$$

$$\leq ||u_o||_{pD} \, M_k \, \frac{M_q^{m+1} \, |||T|||^m}{1 - M_q \, |||T|||} \, . \tag{4.6}$$

Cruder estimates can even be obtained by substituting for $|||T|||$ the bound provided by (4.1).

5. PROBLEMS (i), (ii), (iii) REVISITED

Let us apply the results of Sections 2-4 to the three problems of Section 1. Inasmuch as problems (i) and (ii) have been treated extensively elsewhere [3,4], only a summary of the results pertaining to these problems will appear here.

(i) Problem (i). It was shown in [3] that there exists a unique solution constructed by the method of successive approximation to problem (i) in an open bounded $D \subset R^n$ and belonging to $L_2(D;L_2(\Omega))$ where (Ω,A,P) is the probability space generated by the random data (u_o and q). In particular it was noted that the kernel $\Gamma(x,\xi)$ is an L_2* kernel, i.e. satisfies (4.1-2), if n = 2 or n = 3, in which case the convergence of the method of successive approximations is uniform in D. As a particular example, the problem

$$\frac{d}{dx}\left[k(x;\omega)\frac{d\phi}{dx}\right] = 0, \quad 0 < x < 1$$

$$\phi(0) = 0, \quad \phi(1) = 1, \tag{5.1}$$

$$k(x;\omega) = 1 + \varepsilon(\omega)\, x > 0$$

was considered. Equation (1.8) for this problem takes the form (1.2) with $u = (1+\varepsilon)^{\frac{1}{2}}\phi$, $u_o = x(1+\varepsilon)^{\frac{1}{2}}$, $q(x) = -\varepsilon^2[4(1+\varepsilon x)]^{-2}$, and

$$\Gamma(x,\xi) = \begin{cases} \xi(1-x), & 0 \le \xi \le x \le 1 \\ x(1-\xi), & 0 \le x \le \xi \le 1 \end{cases}$$

is an L_2* kernel.

(ii) Problem (ii) was considered in [4] for $B(D;L_p(\Omega)) = L_2(D;L_2(\Omega))$ and the results were applied to the problem

$$\frac{\partial u}{\partial t} - \frac{\partial^2 u}{\partial x^2} + a(\omega)e^{-kt} = F(x,t;\omega), \quad 0 < x < 1, \ \varepsilon > 0,$$

$$u(0,t;\omega) = u(1,t;\omega) = 0, \quad t > 0, \tag{5.2}$$

$$u(x,0;\omega) = 0, \quad 0 < x < 1.$$

In particular a series expression for $E\{u_m\}$ was obtained for the case where $a \in U(-c,c)$ and is independent of F.

(iii) We now apply the results of this article to problem (iii), more precisely to equation (1.6). We assume the kernel $f(t,\tau)$ is an absolutely integrable function of $t-\tau$ in agreement with the physical background of the problem. We take $B(D) = L_2[0,\infty]$ and $p = p' = 2$. While in general f will not be an L_2 kernel the operator T is nonetheless bounded as we now prove.

For $z\varepsilon L_2[0,\infty]$ we have $Tz = \int_0^t f(t-\tau)z(\tau)d\tau$. Thus

$$|Tz|^2 \leq \int_0^t \int_0^t |f(t-\tau_1)||f(t-\tau_2)|z(\tau_1)\,\bar{z}(\tau_2)d\tau_1 d\tau_2$$

$$\leq \int_0^\infty \int_0^\infty |f(\xi_1)||f(\xi_2)|\; z(t-\xi_1)\,\bar{z}(t-\xi_2)\,d\xi_1\,d\xi_2 \quad .$$

Hence

$$||T_z||^2_{[0,\infty]} = \int_0^\infty |T_z|^2\,dt$$

$$\leq \int_0^\infty \int_0^\infty \int_0^\infty f|(\xi_1)||f(\xi_2)|\; z(t-\xi_1)\,\bar{z}\,(t-\xi_2)\,d\xi_1 d\xi_2 dt$$

$$\leq \{\int_0^\infty |f(\xi)|\,d\xi\}^2\; ||z||^2_{[0,\infty]} < \infty \tag{5.3}$$

since $|\int_0^\infty z(t-\xi_1)\bar{z}(t-\xi_2)\,dt| \leq ||z||^2_{[0,\infty]}$ by Schwarz' inequality and f is absolutely integrable. In fact it can be shown (by considering $z(t) = 1$, $t \leq T$ and 0, $t \geq T$) that

$$|||T||| = \int_0^\infty |f(\xi)|\,d\xi\ . \tag{5.4}$$

Also, for each $z\varepsilon L_2([0,\infty];\,L_2(\Omega))$, and each $t\varepsilon[0,\infty]$, $T[u^*(z)]$ defines a bounded linear functional on $L_2(\Omega)$ since for some $v\varepsilon L_2(\Omega)$ identified with u^*

$$|T[u^*(z)]| = |\int_0^t f(t-\tau)\;(\int_\Omega vz(\tau)dP)d\tau|$$

$$\leq \int_0^t |f(t-\tau)|\int_\Omega |vz(\tau)|dP\;d\tau$$

$$\leq \;||v||_2 \int_0^t |f\;(t-\tau)|\;||z(\tau)||_2\;d\tau$$

$$\leq \;c||v||_2$$

for some constant $c < \infty$. Thus the operator T satisfied the assumptions of Section 2. Consequently there exists a unique solution to equation (1.6) in the sense of Definition 2.1 which can be constructed by the method of successive approximations if (2.5) (or more restrictively (2.7)) is satisfied.

Let us determine the expectation of this solution. From (3.3) we obtain

$$E\{u(t)\} = E\{u_o(t)\} + \sum_{j=1}^{\infty} \int_0^t f(t-\tau_j) \cdots \int_0^{\tau_3} f(\tau_3-\tau_2) \int_0^{\tau_2} f(\tau_2-\tau_1) \cdot$$

$$\cdot E\{q(\tau_1)q(\tau_2) \ldots q(\tau_j)u_o(\tau_1)\} \, d\tau_1 \, d\tau_2 \ldots d\tau_j \,. \tag{5.5}$$

If we truncate the series in (5.5) after the second term we obtain

$$E\{u_2(t)\} = E\{u_o(t)\} + \int_0^t f(t-\tau) \, E\{q(\tau)u_o(\tau)\} d\tau$$
$$+ \int_0^t f(t-\tau) \int_0^{\tau} f(\tau-\xi) E\{q(\tau)q(\xi)u_o(\xi)\} d\xi \, d\tau$$

which when q is (wide sense) stationary and independent of u_o reduces to

$$E\{u_2(t)\} = E\{u_o(t)\} + E\{q\} \int_0^t f(t-\tau) \, E\{u_o(\tau)\} d\tau$$
$$+ \int_0^t f(t-\tau) \int_0^{\tau} f(\tau-\xi) R_q(\tau-\xi) E\{u_o(\xi)\} d\xi \, d\tau, \tag{5.6}$$

where $R_q(\tau-\xi) = E\{q(\tau)q(\xi)\}$ is the correlation function of q.

Approximation (5.6) to (5.5) should be compared to a second order approximation obtained by Distefano [2] using a truncated hierarchy method originally proposed by Richardson [6]. Distefano's approximation is given as the solution to a complicated integral equation (eq. (9.7) in [2]) which may itself have to be solved by some approximate technique. Furthermore, no idea of the "range of goodness" for this approximation can be obtained as is expected from a "dishonest" approximate methods. (We use here Keller's terminology [7]). By contrast, given f, q, and u_o, one can easily evaluate the right hand side of (5.6) and one can always obtain an estimate (possibly quite poor) of the error by using (3.4) or (3.5). These advantages of our method were already pointed out in [3], Section 5, where a comparison between exact, perturbation, and successive approximation solutions to problem (5.1) was carried out.

Our method also allows for an easy asymptotic analysis of the expectation (5.5) in the limit $t \to \infty$. Indeed writing $q = \phi[1+\psi]$ where $\phi = E\{q\}$ and considering (for the sake of comparison of our results with those of [2]) the particular case when $f \geq 0$ and $\psi = \pm \nu$ with equal probability and $E\{u_o\} \to 1$ as $t \to \infty$ we obtain from (5.5)

$$E\{u\} \sim 1 + \phi F + \phi^2(1+\nu^2)F^2 + \phi^3 F^3 + \dots$$

where $F = \int_0^t f(t)\,dt$. This can be written as

$$E\{u\} \sim \frac{1}{2}\{[1-\phi(1+\nu)F]^{-1} + [1-\phi(1-\nu)F]^{-1}\} \quad .$$

Thus we recover Distefano's result (eq. (9.13) in [2]), and in particular his condition (9.15) for the validity of (5.8) is identical with our convergence criterion (2.7) (taking (5.4) into account).

Of course, as pointed out in [5], Section 4.4D, one can define a random kernel $K(t,\tau;\omega) = f(t,\tau)q(\tau;\omega)$ so that equation (1.6) becomes a Volterra integral equation with a random kernel for which there is a sizable body of knowledge (Cf. [5], [8], and the references therein). As a matter of fact one can always define a random operator A by $A(\omega)[u] = T[q(\omega)u]$ so that equation (1.8) now becomes a random operator equation. We believe that one merit of our method is that by not making this transformation to a random operator and by using weak solutions it transforms a random problem (equation (1.8)) into an essentially deterministic problem (equation (2.1)) for which standard methods can be applied. The fact that, in general, only the first and second moments of the solution can be determined is the price we have to pay for this simplification of the problem.

Finally, in spite of the fact that the method of successive approximations is notorious for the slowness of its convergence, our (limited) experience with numerical examples [3,4] tends to indicate a rather fast convergence which compares favorably with other methods [3].

REFERENCES

1. Bécus, G.A. and Cozzarelli, F.A. (1976) SIAM J. Appl. Math., 31, pp. 148-158.
2. Distefano, N. (1968) J. Math. Anal. Appl., 23, 365-383.
3. Bécus, G.A. and Cozzarelli, F.A. (1976) SIAM J. Appl. Math., 31, pp. 159-178.
4. Bécus, G.A. (1978) J. Math. Anal. Appl., 64, in press.
5. Bharucha-Reid, A.T. (1972) Random Integral Equations, Academic Press, New York.
6. Richardson, J.M. (1964) in Proc. Sympos. Appl. Math. XVI, Amer. Math. Soc., Providence, Rhode Island, pp. 290-302.
7. Keller, J.B. (1964) in Proc. Sympos. Appl. Math. XVI, Amer. Math. Soc., Providence, Rhode Island, pp. 145-170.
8. Tsokos, C.P. and Padgett, W.J. (1974) Random Integral Equations with Applications to Life Sciences and Engineering, Academic Press, New York.
9. Karhunen, K. (1947) Ann. Acad. Sci. Fenn., Ser. AI, 37, 1-79.

Applications of the Liouville Equation*

W. E. Boyce

Department of Mathematical Sciences
Rensselaer Polytechnic Institute, Troy, New York 12181

Abstract
1. Introduction
2. The Random Initial Condition Problem
3. A More General Problem
4. Joint Densities, Boundary Value Problems
Acknowledgment
References

*Research supported by U.S. National Science Foundation Grant No. MCS 75-08328.

ABSTRACT

This paper deals with the use of the Liouville equation for the approximate numerical solution of random initial value problems for ordinary differential equations. An algorithm is presented for the determination of the density function of the solution of

$$\frac{du}{dt} = g[u,t,\gamma(t)], \quad u(0) = \alpha ,$$

where α is a given random variable and $\gamma(t)$ is a given stochastic process. Extensions to other problems are also indicated.

1. Introduction. We will be primarily concerned with the random initial value problem

$$\frac{du}{dt} = g[u,t,\gamma(t)], \quad u(0) = \alpha, \tag{1.1}$$

where α is a given random variable and $\gamma(t)$ is a given stochastic process to be specified more fully later. Our object is to determine the marginal (or joint) densities or distributions associated with the solution process $u(t)$ in terms of corresponding quantities for $\gamma(t)$ and α.

One class of methods for handling such problems may be called direct numerical methods; see [2], [3], [4], and [5] for a discussion of several methods of this general type. In all such methods one proceeds "directly" by discretizing the stochastic process $\gamma(t)$ and the random variable α, and reducing the problem in one way or another to the solution of a finite family of deterministic problems. An alternative approach is to consider instead the transition density function of the solution process $u(t)$, and to work with the second order partial differential equation of the diffusion type that it satisfies. This approach has been used frequently and has proven to be extremely valuable; see, for example, [1] and [7], among many other references.

Here we will present a method that, in a sense, is based on a combination of these two viewpoints. It involves the reduction of the given problem (1.1), by a process similar to that used in [4], to a finite family of initial value problems in which randomness enters only through the initial conditions. These problems are then analyzed by means of an associated partial differential equation, the Liouville equation. However, this equation is of first, rather than second, order; and this is a significant simplification.

2. The Random Initial Condition Problem. Let us begin by considering the first order scalar initial value problem for the deterministic differential equation

$$\frac{du}{dt} = g(u,t), \tag{2.1}$$

subject to the random initial condition

$$u(0) = \alpha, \tag{2.2}$$

where g is a given (deterministic) function and α is a random variable with given density $f_\alpha(x)$. If

$$F(x,t) = \Pr\{u(t) \leq x\} \tag{2.3}$$

is the distribution function of u(t), then

$$\Pr\{x_1 < u(t) \leq x_2\} = F(x_2,t) - F(x_1,t) = \int_{x_1}^{x_2} f(x,t)dx , \tag{2.4}$$

where f(x,t) is the associated density function.

It is well known that f(x,t) satisfies the partial differential equation

$$\frac{\partial f}{\partial t} + \frac{\partial}{\partial x}(fg) = 0 \tag{2.5}$$

and the subsidiary conditions

$$f(x,0) = f_\alpha(x), \tag{2.6}$$

$$f(x,t) \to 0 \text{ as } |x| \to \infty \text{ for } t > 0 . \tag{2.7}$$

In probability Eq. (2.5) is known as Liouville's equation, and it can be derived in various ways (see, for example, [7], [8]). It is perhaps worth emphasizing the fact (mentioned in [8]) that this equation also occurs in continuum mechanics, where it is known as the continuity equation, the equation of conservation of mass, or the transport equation. An analogous point of view is possible in probability; as the initial density f_α is swept downstream (see Figure 1) by the process governed by Eq. (2.1), it is transformed according to the partial differential equation (2.5). This amounts to the statement that "probability is conserved." Equation (2.5) can readily be derived by starting with this "conservation law" and proceeding as in the derivation of the continuity equation in fluid mechanics, for instance.

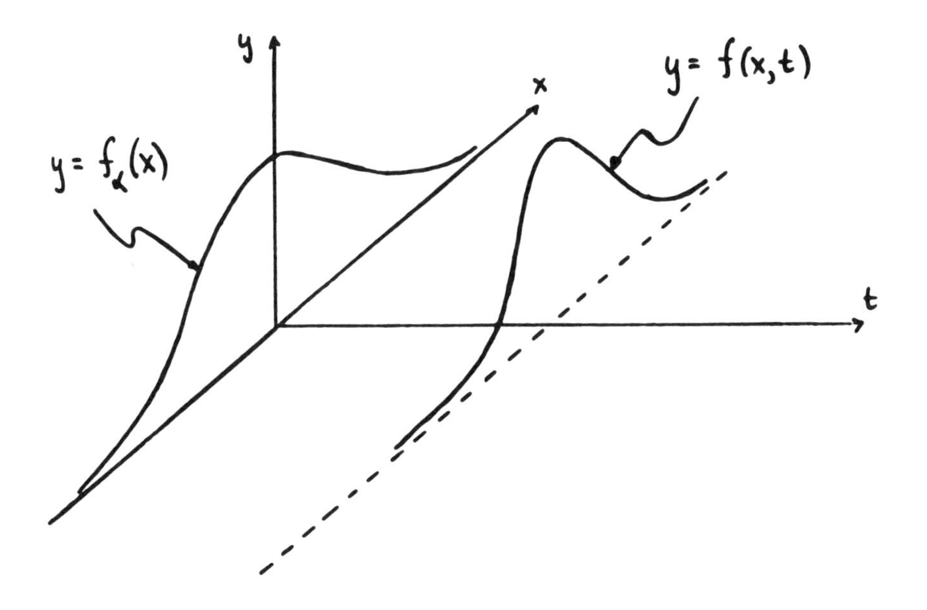

FIGURE 1

Let

$$P(t) = \Pr\{a < u(t) \leq b\} = \int_a^b f(x,t)dx \qquad (2.8)$$

be the probability that u(t) lies in the interval $a < u(t) \leq b$, and let $\Delta P(t,\Delta t)$ be the net change in P in the time interval $(t,t+\Delta t)$. Since "probability is conserved," $\Delta P(t,\Delta t)$ must be due entirely to flow across the boundaries x = a and x = b. Thus

$$\begin{aligned}\Delta P(t,\Delta t) &= -\int_t^{t+\Delta t} f(b,\tau)\frac{du}{d\tau}(b,\tau)d\tau + \int_t^{t+\Delta t} f(a,\tau)\frac{du}{d\tau}(a,\tau)d\tau \\ &= -\int_t^{t+\Delta t} [f(b,\tau)g(b,\tau) - f(a,\tau)g(a,\tau)]d\tau .\end{aligned}$$

Using the mean-value theorem and letting $\Delta t \to 0$, we have

$$\begin{aligned}\frac{dP}{dt} &= -[f(b,t)g(b,t) - f(a,t)g(a,t)] \\ &= -\int_a^b \frac{\partial}{\partial x}[f(x,t)g(x,t)]dx . \end{aligned} \qquad (2.9)$$

On the other hand, from Eq. (2.8), we also have

$$\frac{dP}{dt} = \int_a^b \frac{\partial f}{\partial t}(x,t)dx. \qquad (2.10)$$

Equating the two expressions for dP/dt and invoking the arbitrariness of a and b, we have Liouville's equation (2.5).

Alternatively, one can start with the general physical balance law

$$\frac{\partial Q}{\partial t} + \frac{\partial S}{\partial x} = 0, \qquad (2.11)$$

where Q is a quantity such as mass density and S is the flux of Q per unit time. In the case where the flux is due only to convection, it follows that S is given by Q times velocity, or

$$S = vQ . \qquad (2.12)$$

Thus Eq. (2.11) becomes

$$\frac{\partial Q}{\partial t} + \frac{\partial}{\partial x}(vQ) = 0. \qquad (2.13)$$

By the analogy between mass density and probability density we identify Q with f; also v is identified with du/dt, or g. Hence Eq. (2.13) becomes the Liouville equation (2.5).

In the case where the flux also has a diffusive component, Eq. (2.12) is replaced by

$$S = vQ - \sigma\frac{\partial Q}{\partial x}, \tag{2.14}$$

where $\sigma \geq 0$ is the coefficient of diffusion. The minus sign occurs because diffusion results in a flow from regions of high concentration (large Q) to those of small concentration. Substituting Eq. (2.14) in Eq. (2.11), we obtain the equation

$$\frac{\partial Q}{\partial t} + \frac{\partial}{\partial x}(vQ) - \frac{\partial}{\partial x}\left(\sigma\frac{\partial Q}{\partial x}\right) = 0 \tag{2.15}$$

which has the form of the forward Kolmogorov equation.

The generalization to systems of arbitrary order is immediate. Suppose that

$$\frac{d\underset{\sim}{u}}{dt} = \underset{\sim}{g}(\underset{\sim}{u},t), \quad \underset{\sim}{u}(0) = \underset{\sim}{\alpha}, \tag{2.16}$$

where

$$\underset{\sim}{u} = \begin{pmatrix} u_1 \\ \vdots \\ u_n \end{pmatrix} \quad \underset{\sim}{\alpha} = \begin{pmatrix} \alpha_1 \\ \vdots \\ \alpha_n \end{pmatrix} \quad \underset{\sim}{g}(\underset{\sim}{u},t) = \begin{pmatrix} g_1(u_1,\ldots,u_n,t) \\ \vdots \\ g_n(u_1,\ldots,u_n,t) \end{pmatrix}. \tag{2.17}$$

Let $f(\underset{\sim}{x},t) = f(x_1,\ldots,x_n,t)$ be the joint density of $u_1(t),\ldots,u_n(t)$, and denote the joint density of $\alpha_1,\ldots,\alpha_n$ by $f_{\underset{\sim}{\alpha}}(\underset{\sim}{x})$. Then $f(\underset{\sim}{x},t)$ satisfies the partial differential equation

$$\frac{\partial f}{\partial t} + \sum_{i=1}^{n} \frac{\partial}{\partial x_i}(g_i f) = 0 \tag{2.18}$$

with the subsidiary conditions

$$f(\underset{\sim}{x},0) = f_{\underset{\sim}{\alpha}}(\underset{\sim}{x}), \tag{2.19}$$

$$f(\underset{\sim}{x},t) \to 0 \text{ as } |x| \to \infty. \tag{2.20}$$

The solution of the problems (2.5), (2.6), (2.7) can be obtained in several ways. It will be convenient for our purposes to express it in terms of characteristics. First rewrite Eq. (2.5) as

$$\frac{\partial f}{\partial t} + g\frac{\partial f}{\partial x} = -\frac{\partial g}{\partial x}f \ . \tag{2.21}$$

If we choose the curve $t = t(\sigma)$, $x = x(\sigma)$ to satisfy the characteristic equations

$$\frac{dt}{d\sigma} = 1, \ \frac{dx}{d\sigma} = g[x(\sigma), \ t(\sigma)] \ , \tag{2.22}$$

then on this characteristic curve, Eq. (2.21) reduces to the ordinary differential equation

$$\frac{df}{d\sigma} = \hat{g}(\sigma)f \ , \tag{2.23}$$

where

$$\hat{g}(\sigma) = -\frac{\partial g}{\partial x}[x(\sigma), \ t(\sigma)] \ . \tag{2.24}$$

Consequently

$$f(x,t) = f(x_0,0) \exp \int_0^t \hat{g}(\sigma)d\sigma \ , \tag{2.25}$$

where $(x_0,0)$ is the point where the characteristic curve through (x,t) crosses the x-axis. This method of solution also generalizes at once to higher dimensions.

The case in which a random vector appears in the differential equation can be handled in a similar fashion, as shown in [7]. An application of this situation appears in [6].

3. A More General Problem. We now turn to the more interesting problem in which the differential equation depends on a stochastic process. For simplicity, we consider a first order scalar equation

$$\frac{du}{dt} = g[u,t,\gamma(t)] \ , \tag{3.1}$$

where $\gamma(t)$ is a given stochastic process, subject again to the random initial condition

$$u(0) = \alpha \ . \tag{3.2}$$

The basic idea is to reduce this problem to a sequence of initial value problems of the type already considered, that is, problems in which randomness enters only through the initial conditions. We first assume that γ is of a special class that makes this reduction possible in a straightforward way.

We establish a mesh on t

$$\{0,t_1,t_2,\ldots,t_i,t_{i+1},\ldots\},\ \Delta t_i = t_{i+1} - t_i \tag{3.3}$$

and an x

$$\{\ldots,x_{-1},x_0,x_1,x_2,\ldots,x_j,\ldots\} . \tag{3.4}$$

Let

$$\gamma_i = \gamma(t_i),\ \Delta\gamma_i = \gamma_{i+1} - \gamma_i . \tag{3.5}$$

Then we make the following assumptions about $\gamma(t)$.

(a) There is a number δ and a positive integer P such that $\gamma_i = p\delta$ for each i, where p is an integer and $-P \le p \le P$. Consequently, $\Delta\gamma_i = q\delta$ for each i, where q is an integer and $-Q \le q \le Q$ for some positive integer Q.

(b) $\gamma(t)$ is linear in each subinterval. Thus for $t_i \le t \le t_{i+1}$ we can write

$$\gamma(t) = \gamma_{pq}(t) = p\delta + \frac{t-t_i}{\Delta t_i} q\delta . \tag{3.6}$$

(c) The joint probability

$$\rho_i(p,q) = \Pr\{\gamma_i = p\delta, \Delta\gamma_i = q\delta\} \tag{3.7}$$

is known.

It is clear that stochastic processes satisfying these rather stringent requirements belong to a special class. However, the usefulness of this approach is enhanced by the observation that many other stochastic processes can be approximated arbitrarily well by processes whose sample functions satisfy requirements (a), (b), and (c). For example, it is shown in [4] that every mean square continuous process with independent increments can be so approximated.

We will now describe an algorithm by which the density function f(x,t) for the initial value problem (3.1), (3.2) can be found when $\gamma(t)$ satisfies conditions (a), (b), and (c). We assume that the value of $f(x_j,t_i) = f_{ji}$ for each j has been found by prior calculations, and will describe how to make the transition to the next time step at $t = t_{i+1}$. Assuming that $\rho_i(p,q)$ is known [condition (c)] and fixing p and q, we determine $\gamma_{pq}(t)$ by condition (b). This in turn determines

$$g_{pq}(x,t) = g[x,t,\gamma_{pq}(t)], \quad t_i \leq t \leq t_{i+1} . \tag{3.8}$$

Next we find the characteristic through (x_j,t_i) by solving the initial value problem

$$\frac{dx}{dt} = g_{pq}(x,t), \quad x(t_i) = x_j . \tag{3.9}$$

Denoting the solution of this problem by $x = x_j(t)$, we let $\xi_j = x_j(t_{i+1})$. In other words, ξ_j is the point where the characteristic through (x_j,t_i) crosses the line $t = t_{i+1}$.

Now we propagate f along the characteristics that have just been found by solving the initial value problems

$$\frac{df}{dt} = -g_1[x_j(t),t,\gamma_{pq}(t)]f = \hat{g}(t)f , \tag{3.10}$$

$$f(x_j,t_i) = f_{ji} . \tag{3.11}$$

The solution of this problem can be expressed in the form

$$f[x_j(t),t] = f_{ji} \exp \int_{t_i}^{t} \hat{g}(\tau)d\tau, \quad t_i \leq t \leq t_{i+1} \tag{3.12}$$

Evaluating $f[x_j(t),t]$ when $t = t_{i+1}$, we obtain

$$f_{pq}(\xi_j,t_{i+1}) = f_{ji} \exp \int_{t_i}^{t_{i+1}} \hat{g}_{pq}(\tau)d\tau . \tag{3.13}$$

In Eq. (3.13) the subscripts are inserted to remind us that this solution depends on the initial choice of p and q. Note that Eq. (3.13) gives f_{pq} at the time step t_{i+1} at the points $\{\xi_j\}$ and not at the original mesh points $\{x_j\}$. Thus an interpolation procedure must be employed to deduce $\{f_{pq}(x_j,t_{i+1})\}$ from $\{f_{pq}(\xi_j,t_{i+1})\}$.

Finally, we must repeat the foregoing procedure for each allowable pair of values for p and q. Then we obtain the final result from the linear combination

$$f(x_j,t_{i+1}) = \sum_q \sum_p \rho_i(p,q) f_{pq}(x_j,t_{i+1}) . \tag{3.14}$$

Under rather mild conditions on g the integration of the initial value problem (3.9) can be carried out with any desired degree of accuracy by choosing sufficiently many steps to cover the interval (t_i,t_{i+1}). Similarly, the interpolation process that generates $\{f_{pq}(x_j,t_{i+1})\}$ from $\{f_{pq}(\xi_j,t_{i+1})\}$ can be made arbitrarily accurate by choosing a sufficiently dense mesh $\{x_j\}$. Thus, in principle, it is possible to calculate $f(x_j,t_{i+1})$ from Eq. (3.14) with any specified degree of accuracy under the conditions that have been enumerated. Of course, this analysis neglects the effect of round-off error. Also, if $\gamma(t)$ does not satisfy conditions (a), (b), and (c), but is replaced by an approximating stochastic process that does go, then it will also be necessary to specify a sufficiently fine mesh on the t-axis.

Note that condition (b) is hardly crucial, that is, it is not essential for $\gamma(t)$ to be piecewise linear. An attractive alternative would be to assume that the sample functions of γ are constant in each subinterval with jumps occurring at the points t_i. Or we could assume them to be piecewise quadratic, or to have some other given form.

4. Joint Densities, Boundary Value Problems. The approach described here can also be used for various joint densities of second or higher order, as well as for two-point boundary value problems. In the latter case, the boundary conditions render the problem significantly more difficult.

Consider the second order system

$$\frac{du_1}{dt} = g_1(u_1,u_2,t) \; , \qquad \frac{du_2}{dt} = g_2(u_1,u_2,t) \; , \tag{4.1}$$

subject to the random boundary conditions

$$u_1(0) = \alpha, \quad u_1(1) = \beta \; , \tag{4.2}$$

where α and β are independent random variables with given densities f_α and f_β respectively. Let $f(x_1,x_2,t)$ be the joint density of $u_1(t)$ and $u_2(t)$. Then the Liouville equation for f is

$$\frac{\partial f}{\partial t} + \frac{\partial}{\partial x_1}(g_1 f) + \frac{\partial}{\partial x_2}(g_2 f) = 0 \qquad (4.3)$$

on the domain $-\infty < x_1, x_2 < \infty$, $0 \leq t \leq 1$. The boundary conditions corresponding to (4.2) are

$$\int_{-\infty}^{\infty} f(x_1,x_2,0)\,dx_2 = f_\alpha(x_1) \,, \qquad (4.4)$$

$$\int_{-\infty}^{\infty} f(x_1,x_2,1)\,dx_2 = f_\beta(x_1) \,, \qquad (4.5)$$

and in addition we must have

$$f(x_1,x_2,t) \to 0 \text{ as } |x_1|,\ |x_2| \to \infty \,. \qquad (4.6)$$

Because of the integral boundary conditions it is not obvious how to solve the indicated problem for f. Worse, even if a solution method were available, the extension to boundary value problems in which a stochastic process appears in Eq. (4.1) seems not to follow as in the case of initial value problems.

Acknowledgment. Some of the ideas expressed here had their origin in conversations with Werner E. Kohler, now of Virginia Polytechnic Institute and State University.

REFERENCES

1. L. Arnold, Stochastic differential equations: theory and applications, John Wiley and Sons, Inc., 1974.

2. M. R. Barry and W. E. Boyce, Numerical Solution of a Class of Random Boundary Value Problems, J. Math. Anal. Appl. (to appear).

3. W. E. Boyce, Approximate Solution of Random Ordinary Differential Equations, Adv. Appl. Prob. 10, 172-184 (1978).

4. W. E. Kohler and W. E. Boyce, A Numerical Analysis of some First Order Stochastic Initial Value Problems, SIAM J. Appl. Math. 27, 167-179 (1974).

5. M. D. Lax and W. E. Boyce, The Method of Moments for Linear Random Initial Value Problems, J. Math. Anal. Appl. 53, 111-132 (1976).

6. W. J. Padgett, G. Schultz, and C. P. Tsokos, A Random Differential Equation Approach to the Probability Distribution of BOD and DO in Streams, SIAM J. Appl. Math. 32, 467-483 (1977).

7. T. T. Soong, Random Differential Equations in Science and Engineering, Academic Press, 1973.

8. R. Syski, Stochastic Differential Equations, Chapter 8 in Modern Nonlinear Equations by T. L. Saaty, McGraw-Hill, Inc., 1967.

Approximate Solution of Fixed Point Equations for Random Operators*

P. S. Chandrasekharan and A. T. Bharucha-Reid

Department of Mathematics
Wayne State University, Detroit, Michigan 48202

1. Introduction
2. Some Results from Probabilistic Operator Theory
3. Existence, Uniqueness, and Convergence of the Approximate Random Solutions
References

*Research supported by U.S. Army Research Office Grant No. DAAG29-77-G-0164.

1. Introduction

Let $\mathcal{X}$ be a Banach space, and let T be an operator mapping $\mathcal{X}$ into itself. An element $x \in \mathcal{X}$ such that $Tx = x$ is said to be a *fixed point* of T. It is clear that the set of fixed points of an operator T is the same as the solution set of the eigenvalue problem $(T - \lambda I)x = 0$, corresponding to the eigenvalue $\lambda = 1$. Consider a concrete operator equation $Sx = y$ in a given Banach space. Then, finding a fixed point of T is equivalent to obtaining a solution of the operator equation $Tx \equiv Sx - x - y$. Fixed point theorems, and methods based on fixed point theorems, play a central role in the theory of operator equations.

In this paper we give a probabilistic generalization of a theorem of R. Weiss [9] on the approximation of fixed points of nonlinear compact operators. Weiss was concerned with the approximate solution of operator equations of the form

$$Tx - x = 0 , \tag{1.1}$$

where T is a compact operator defined on a domain $\mathcal{D}$ of a Banach space $\mathcal{X}$ into $\mathcal{X}$. In particular, he was concerned with the development of numerical procedures for the solution of Eq. (1.1) when Eq. (1.1) is replaced by the family of equation

$$T_n x - x = 0 , \qquad n \geq 1 , \tag{1.2}$$

where $\{T_n, n \geq 1\}$ is a collectively compact family, and the T_n converge pointwise to T. We recall (Anselone [1]) that a family of operators $\{T_n\}$ is said to be *collectively compact* if the operators T_n map the unit ball of $\mathcal{X}$ into one and the same compact subset; that is, $\overline{\bigcup_n T_n[B]}$ is compact, where B denotes the unit ball of $\mathcal{X}$.

Weiss established the existence, uniqueness, and convergence of solutions x_n of Eq. (1.2) in a neighborhood of an isolated solution* of Eq. (1.1).

In Sect. 2 we give some definitions and theorems from probabilistic operator theory which will be used in Sect. 3, where we state and prove a probabilistic generalization of Weiss' theorem.

---
*An element $x \in \mathcal{X}$ is said to be an *isolated solution* of Eq. (1.1) if $[I - T'(x)]$ is nonsingular.

2. Some Results from Probabilistic Operator Theory

Let $(\Omega, \mathcal{A}, \mu)$ be a complete probability space; and let $(\mathcal{X}, \mathcal{B})$ and $(\mathcal{Y}, \mathcal{C})$ be measurable spaces, where $\mathcal{X}$ and $\mathcal{Y}$ are separable Banach spaces and $\mathcal{B}$ and $\mathcal{C}$ are the σ-algebras of Borel subsets of $\mathcal{X}$ and $\mathcal{Y}$, respectively.

Definition 1. A mapping $x\colon \Omega \to \mathcal{X}$ is said to be an $\mathcal{X}$-valued random variable (random element or generalized random variable) if the inverse image under the mapping x of every $B \in \mathcal{B}$ belongs to $\mathcal{A}$; that is, $x^{-1}(B) \in \mathcal{A}$ for all $B \in \mathcal{B}$.

Let T be an operator on $\mathcal{X}$ into $\mathcal{Y}$ with domain $\mathcal{D}(T)$.

Definition 2. A mapping $T\colon \Omega \times \mathcal{D}(T) \to \mathcal{Y}$ is said to be a random operator if $T(\omega)x = y(\omega)$ is a $\mathcal{Y}$-valued random variable for every $x \in \mathcal{D}(T)$. (We remark that, in general, $\mathcal{D}(T)$ will depend on ω).

Definition 3. A random operator $T(\omega)$ on $\mathcal{X}$ is said to be (a) linear if $T(\omega)[\alpha x_1 + \beta x_2] = \alpha T(\omega)x_1 + \beta T(\omega)x_2$ almost surely (a.s.) for all $x_1, x_2 \in \mathcal{D}(T)$, α, β scalars; and (b) bounded if there exists a nonnegative real-valued random variable $M(\omega)$ such that for $x \in \mathcal{D}(T)$, $\|T(\omega)x\| \le M(\omega)\|x\|$ a.s.

Definition 4. A random operator $T(\omega)$ on $\mathcal{X}$ is said to be continuous at x_o if $\lim_{n\to\infty} \|x_n - x_o\| = 0$ implies $\lim_{n\to\infty} \|T(\omega)x_n - T(\omega)x_o\| = 0$ a.s.

Let $\mathcal{L}(\mathcal{X})$ denote the Banach algebra of all endomorphisms of $\mathcal{X}$ (i.e., the algebra of all bounded linear operators on $\mathcal{X}$); and let $\mathcal{T}$ denote the σ-algebra of Borel subsets of $\mathcal{L}(\mathcal{X})$, provided the norm topology is assumed.

Definition 5. A mapping $T\colon \Omega \to \mathcal{L}(\mathcal{X})$ is said to be a random endomorphism of $\mathcal{X}$ if $T(\omega)$ is an $\mathcal{L}(\mathcal{X})$-valued random variable; that is, $T^{-1}(F) \in \mathcal{A}$ for every $F \in \mathcal{T}$.

We now introduce the notion of a separable random operator (cf. [2]), a notion of increasing usefulness in probabilistic operator theory (cf. Kannan and Salehi [7], Nashed and Salehi [8]).

Definition 6. A random operator $T(\omega)$ on a Banach space $\mathcal{X}$ is said to be separable if there exists a countable dense set S in $\mathcal{X}$ and a negligible set $N \in \mathcal{O}$ (i.e., $\mu(N) = 0$) such that

$$\{\omega : T(\omega)x \in K,\ x \in F\} \,\Delta\, \{\omega : T(\omega)x \in K,\ x \in F \cap S\} \subset N$$

for every compact set K and every open set F.

We now consider random operators.

Definition 7. An operator equation of the form

$$T(\omega)x = y(\omega) \,, \tag{2.1}$$

where $T(\omega)$ is a given random operator on $\mathcal{X}$ to itself, and $y(\omega)$ is a given $\mathcal{X}$-valued random variable, is said to be a random operator equation; and any $\mathcal{X}$-valued random variable $x(\omega)$ which satisfies the condition

$$\mu(\{\omega : T(\omega)x(\omega) = y(\omega)\}) = 1 \tag{2.2}$$

is said to be a random solution of Eq. (2.1).

In this paper we will make use of the following results on random contraction mappings and inverses of random operators:

(1) Let $T(\omega)$ be a continuous random operator from $\Omega \times \mathcal{X}$ into $\mathcal{X}$, and let $k(\omega)$ be a nonnegative real-valued random variable such that

$$\mu(\{\omega : k(\omega) < 1\}) = 1 \,.$$

Definition 8. A random operator $T(\omega)$ on $\mathcal{X}$ is said to be a random contraction operator if

$$\mu(\{\omega : \|T(\omega)x - T(\omega)y\| \leq k(\omega)\|x-y\|\}) = 1 \tag{2.3}$$

for every $x, y \in \mathcal{X}$.

Now, if $T(\omega)$ is a random contraction operator on $\mathcal{X}$, there exists an $\mathcal{X}$-valued random variable $x(\omega)$ such that $T(\omega)x(\omega) = x(\omega)$ a.s. (cf. Bharucha-Reid [2,3] and Hanš [5]).

(2) If $T(\omega)$ is an $\mathcal{L}(\mathcal{X})$-valued random variable, then $T^{-1}(\omega)$ is the $\mathcal{L}(\mathcal{X})$-valued random variable which maps $T(\omega)x$ into x a.s.

Definition 9. $T(\omega)$ is said to be invertible if $T^{-1}(\omega)$ exists.

Now, let $T(\omega)$ be an invertible random operator with values in $\mathcal{L}(\mathcal{X})$. Then $T^{-1}(\omega)$ is a random operator with values in $\mathcal{L}(\mathcal{X})$ (cf. Hanš [5]).

(3) Let $T(\omega)$ be a separable random operator on $\mathcal{X}$ into itself such that a.s. $T(\omega)$ is invertible and its inverse $T^{-1}(\omega)$ is continuous. Then $T^{-1}(\omega)$ is also a random operator from $\mathcal{X}$ into itself (cf. Nashed and Salehi [8]).

Finally, we define the Fréchet derivative of a random operator.

Definition 10. Let $T(\omega)$ be a random operator on $\mathcal{X}$ into itself such that for every $\omega \in \Omega$ the operator $T(\omega): \mathcal{X} \to \mathcal{X}$ is differentiable; that is, for a fixed $\omega \in \Omega$ and for a given $x_o \in \mathcal{X}$, the limit

$$\lim_{t \to 0} \frac{T(\omega)(x_o + t) - T(\omega)x_o}{t}, \quad t > 0, \tag{2.4}$$

exists. We denote this $\mathcal{X}$-valued random element by $T'(\omega)(x_o)$. The randomness of $T(\omega)$ implies that its Fréchet derivative at x_o, i.e., $T'(\omega)(x_o) : \Omega \times \mathcal{X} \to \mathcal{X}$, is random (cf. Bharucha-Reid and Kannan [4]).

3. Existence, Uniqueness and Convergence of the Approximate Random Solutions.

Given a complete probability measure space $(\Omega, \mathcal{A}, \mu)$ and a separable Banach space $(\mathcal{X}, \mathcal{B})$, we wish to consider the solution of the fixed point equation

$$T(\omega)x - x = 0, \tag{3.1}$$

where $T(\omega)$ is a compact random operator on $\mathcal{X}$ to itself. In particular, we will consider the family of fixed point equations

$$T_n(\omega)x - x = 0. \tag{3.2}$$

We make the following assumptions:

(i) $\{T_n(\omega),\ n \geq 1\}$ is an a.s. collectively compact family, with $\mathcal{D}(T_n(\omega)) = \mathcal{D} \subset \mathfrak{X}$;

(ii) $T_n(\omega) \to T(\omega)$ strongly a.s. on $\mathcal{D}$;

(iii) The operators $T_n(\omega)$ possess continuous first and bounded second (Fréchet) derivatives on $S(\tilde{x}(\omega),r)$ - a ball of radius r with center $\tilde{x}(\omega)$, where $T(\omega)\tilde{x}(\omega) = \tilde{x}(\omega)$ a.s. ; that is, $\tilde{x}(\omega)$ is a random fixed point of $T(\omega)$.

We want to show that if $\tilde{x}(\omega)$ is an isolated solution of Eq.(3.1), then $\tilde{x}(\omega)$ is a random solution. In order to do this we will construct a sequence of approximate solutions $x_n(\omega)$ such that $x_n(\omega) \to \tilde{x}(\omega)$ a.s., and where each $x_n(\omega)$ is a random solution of Eq.(3.2).

The following lemma is immediate, using an argument identical to that of Weiss, and the results of Hanš or Nashed and Salehi given in Sect. 2.

Lemma. Let the assumptions (i) - (iii) hold, and assume that $[I - T'(\omega)(\tilde{x}(\omega))]$ is nonsingular a.s. Then, the linear random operators $[I - T_n'(\omega)(\tilde{x}(\omega))]$ are nonsingular a.s., and

$$(3.3)\quad \|[I - T_n'(\omega)(\tilde{x}(\omega))]^{-1}\| \leq \beta < \infty\ , \quad n \geq n_1$$

almost surely.

We now state and prove the following result:

Theorem. Let the assumptions (i) - (iii) hold, and assume that $[I - T'(\omega)(\tilde{x}(\omega))]$ is nonsingular a.s. ; that is, $\tilde{x}(\omega)$ is an isolated solution of Eq.(3.1) . Then, there exists a constant $\rho\ (o < \rho \leq r)$ such that for n sufficiently large, Eq. (3.2) has a unique random solution $x_n(\omega) \in S(\tilde{x}(\omega),\ \rho)$ and $\|x_n(\omega) - \tilde{x}(\omega)\| \to 0$ a.s. as $n \to \infty$.

Proof. From the lemma, for $n \geq n_1$, Eq. (3.2) can be rewritten as

$$(3.4)\quad x(\omega) = x(\omega) - [I - T_n'(\omega)(\tilde{x}(\omega))]^{-1}\ [x(\omega) - T_n(\omega)x(\omega) \equiv \Gamma_n(\omega)x(\omega)\ .$$

For any $\xi_1(\omega)\ ,\ \xi_2(\omega) \in S(\tilde{x}(\omega),\ \rho),\ \rho \leq r$, we have

(3.5) $\Gamma_n(\omega)\xi_1(\omega) - \Gamma_n(\omega)\xi_2(\omega)$

$$= [\xi_1(\omega) - \xi_2(\omega)] - [I - T_n'(\omega)(\tilde{x}(\omega))]^{-1}$$

$$\cdot [\xi_1(\omega) - \xi_2(\omega) - (T_n(\omega)\xi_1(\omega) - T_n(\omega)\xi_2(\omega))]$$

$$= [I - T_n'(\omega)(\tilde{x}(\omega))]^{-1} \cdot [\gamma_n(\omega) - T_n'(\tilde{x}(\omega))] \cdot [\xi_1(\omega) - \xi_2(\omega)] ,$$

where

(3.6) $$\gamma_n(\omega) = \int_0^1 T_n'(\omega)\ (\lambda\xi_1(\omega) + (1-\lambda)\xi_2(\omega))d\lambda .$$

We must establish that $\gamma_n(\omega)x$ is an $\mathfrak{X}$-valued random variable, where $x \in \mathfrak{X}$. The integral in (3.6) will be regarded as a Bochner integral; hence we must show that the integrand is Bochner integrable.

Let

$$\tau_n(\omega) = T_n'(\omega)(\lambda\xi_1(\omega) + (1-\lambda)\xi_2) .$$

Firstly, $\lambda\xi_1(\omega) + (1-\lambda)\xi_2(\omega)$ is an $\mathfrak{X}$-valued random variable since $\mathfrak{X}$ is separable. Indeed, $\lambda\xi_1(\omega) + (1-\lambda)\xi_2(\omega)$ is in $S(\tilde{x}(\omega), \rho)$ since $S(\tilde{x}(\omega), \rho)$ is convex. Secondly, we know (cf. [4]) that $T_n'(\omega)$, for every n $(n \geq n_1)$ is a random linear operator. Hence $T_n'(\omega)x$ is an $\mathfrak{X}$-valued random variable; in particular, with values in $S(\tilde{x}(\omega), \rho)$. Since $\mathfrak{X}$ is separable, $\tau_n(\omega)x$ is strongly measurable. Hence, in order to establish the Bochner integrability of $\tau_n(\omega)$ we must show that $\int_\Omega \|\tau_n(\omega)x\|\, d\mu(\omega) < \infty$ (cf. Hille and Phillips [6], p.80). That this condition is satisfied is obvious; hence $\tau_n(\omega)x$ is Bochner integrable, and $\gamma_n(\omega)x$ is a well-defined $\mathfrak{X}$-valued random variable for every $n \geq n_1$.

From the boundedness of $T_n''(\omega)$ (whose bound we denote by α), we have

$$\|\gamma_n(\omega) - T_n'(\omega)(\tilde{x}(\omega))\|$$

$$= \|\int_0^1 T_n'(\omega)(\lambda\xi_1(\omega) + (1-\lambda)\xi_2(\omega))d\lambda$$

$$- T_n'(\omega)(\tilde{x}(\omega))\| \leq \alpha\rho .$$

Hence, from (3.5), using (3.3), we have

(3.7) $$\|\Gamma_n(\omega)\xi_1(\omega) - \Gamma_n(\omega)\xi_2(\omega)\| \leq k\ \|\xi_1(\omega)-\xi_2(\omega)\| ,$$

where $k = \alpha\beta\rho$; and ρ is selected small enough so that $k \leq 1$.

From (3.4) we can write

$$\Gamma_n(\omega)\tilde{x}(\omega) - \tilde{x}(\omega) = [I - T_n'(\omega)(\tilde{x}(\omega))]^{-1} \cdot [T_n(\omega)\tilde{x}(\omega) - \tilde{x}(\omega)] .$$

Hence, from assumption (ii), for n sufficiently large (say $n \geq n_2$) we have

$$\| \Gamma_n(\omega)\tilde{x}(\omega) - \tilde{x}(\omega)\| \leq (1-k)\rho .$$

Therefore, if $n \geq n_3$, where $n_3 = \max\{n_1,n_2\}$, we have for $x(\omega) \in S(\tilde{x}(\omega),\rho)$

$$\| \Gamma_n(\omega)x(\omega) - \tilde{x}(\omega)\| \leq \| \Gamma_n(\omega)x(\omega) - \Gamma_n(\omega)\tilde{x}(\omega)\| + \| \Gamma_n(\omega)\tilde{x}(\omega) - \tilde{x}(\omega)\|$$

$$\leq k\rho + (1-k)\rho = \rho .$$

Hence, for $n \geq n_3$, $\Gamma_n(\omega)$ maps $S(\tilde{x}(\omega),\rho)$ into itself; and by (3.7) $\Gamma_n(\omega)$ is a random contraction mapping on $S(\tilde{x}(\omega),\rho)$. This establishes the existence, uniqueness and measurability of the approximate solutions for each n.

To establish convergence of the approximate random solutions, we can subtract the equation $T_n(\omega)x_n(\omega) - x_n(\omega) = 0$ from the equation $T(\omega)x(\omega) - x(\omega) = 0$, and obtain

$$[I - T_n'(\omega)(\tilde{x}(\omega))] \cdot [x_n(\omega) - \tilde{x}(\omega)] = T_n(\omega)\tilde{x}(\omega) - T(\omega)\tilde{x}(\omega)$$

$$- [T_n(\omega)\tilde{x}(\omega) - T_n(\omega)x(\omega) - T_n'(\omega)(\tilde{x}(\omega))] \cdot [\tilde{x}(\omega) - x_n(\omega)] .$$

Using assumption (iii) and the lemma, we have

$$\|x_n(\omega) - \tilde{x}(\omega)\| \leq \beta \{\|T_n(\omega)\tilde{x}(\omega) - T(\omega)\tilde{x}(\omega)\| + \frac{1}{2}\alpha\|x_n(\omega) - \tilde{x}(\omega)\|\} .$$

Hence

$$\|x_n(\omega) - \tilde{x}(\omega)\| \leq \frac{\beta\|T_n(\omega)\tilde{x}(\omega) - T(\omega)\tilde{x}(\omega)\|}{(1-\frac{\alpha\beta}{2}\|x_n(\omega)-\tilde{x}(\omega)\|} \leq \frac{\beta}{1-k/2}\|T_n(\omega)\tilde{x}(\omega) - T(\omega)\tilde{x}(\omega)\| .$$

Convergence follows from assumption (ii); and the uniqueness and measurability of the random solution $\tilde{x}(\omega)$ follows from the uniqueness and measurability of the approximate solutions $x_n(\omega)$.

References

[1] Anselone, P.M., Collectively Compact Operator Approximation Theory, Prentice-Hall, Englewood Cliffs, N.J., 1971.

[2] Bharucha-Reid, A.T., Random Integral Equations, Academic Press, New York, 1972.

[3] Bharucha-Reid, A.T., Fixed point theorems in probabilistic analysis, Bull. Amer. Math. Soc. 82 (1976), 641-657.

[4] Bharucha-Reid, A.T. and Kannan, R., Newton's method for random operator equations, to appear.

[5] Hans, O., Random operator equations, Proc. 4th Berkeley Sympos. on Math, Statist. and Probability (1960), Vol. II, Univ. of California Press, Berkeley, Calif., 1961, pp. 185-202.

[6] Hille, E. and Phillips, R.S., Functional Analysis and Semi-Groups, American Mathematical Society, Providence, Rhode Island, 1957.

[7] Kannan, R. and Salehi, H., Mésurabilité du point fixé d'une tranformation aléatoire séparable, C.R. Acad. Sci. Paris Sér. A-B 281 (1975), A663-A664.

[8] Nashed, M.Z. and Salehi, H., Measurability of generalized inverses of random linear operators, SIAM J. Appl. Math. 25 (1973), 681-692.

[9] Weiss, R., On the approximation of fixed points of nonlinear compact operators, SIAM J. Numer. Anal. 11 (1974), 550-553.

Approximate Solution of Random Evolution Equations*

P.-L. Chow

Department of Mathematics
Wayne State University, Detroit, Michigan 48202

1. Introduction
2. Approximation of Linear Random Evolution Equations
3. Approximate Moments for Bilinear Random Equations
4. Stochastic Modeling and Other Methods

References

*The preparation of this paper was supported by the U.S. Army Research Office under Grant DAAG-29-78-G-0042 and by NASA under Grant NSG-1330.

1. Introduction

Many initial-boundary value problems for partial differential equations can be regarded as an abstract evolution equation in function space. For examples of random evolution equations, one is referred to [1] and [2]. The evolutional equation approach has two obvious advantages, the conceptual simplication of problems and the notational ecomony.

In the present article, we shall discuss a few ideas in the approximate solution of random evolution equations. In Section 2, linear random evolution equations of Ito type will be treated. By formulating the problem as an initial-value problem in the abstract Wiener space, the method of projection based on the eigenfunction expansions, or more generally, the orthogonal residual is used to obtain an approximate solution. Error estimates for various approximations are explicitly given in terms of the spectral property of the deterministic operator. Section 3 is concerned with bilinear random equations, or evolution equations with random coefficients. Here we are specifically interested in approximating the random bilinear equation by one that has closed moment equations. If the coefficient is a Wiener functional that admits a stochastic integral representation, a reasonable approximation can be achieved by a method of weighted residual. In general it is taken as a formal approximation without mathematical justification. Finally in Section 4, the stochastic modeling technique of replacing the original problem by a solvable one is briefly discussed. In particular, the technique of random coupling is reviewed in the light of a statistical estimation concept. Other practical methods are also mentioned.

2. Approximation of Linear Random Evolution Equations

It is instructive to first consider a concrete example. What we have in mind is the initial-boundary value problem for the heat equation in one dimension

$$\frac{\partial u}{\partial t} - \frac{\partial^2 u}{\partial x^2} = \dot{W}(t,x) \ , \ t > 0 \ , \ x \in (0, \pi) \ . \tag{2.1}$$

$$u(t,0) = u(t,\pi) = 0 \tag{2.2}$$

$$u(0,x) = \eta(x) . \tag{2.3}$$

where $\dot{W}_t \equiv \dot{W}(t, \cdot)$ is a "space-time" white noise so that W_t will be called a Brownian motion, and the function η is given. Let $H = L^2[o,\pi]$, $\mathcal{C}_o^2[o,\pi]$ be the set of all $\mathcal{C}^2$-functions on $[o,\pi]$ that vanish at the end points. Define $A_o = -\frac{d^2}{dx^2}$ restrictive to act on $\mathcal{C}_o^2$, and consider the eigenvalue problem

$$A_o\phi = \lambda \phi . \tag{2.4}$$

Clearly $\lambda_n = n^2$ and $e_n(x) = \sqrt{\frac{2}{\pi}} \sin nx$, $n = 1,2, \ldots$, are the eigenvalues and the normalized eigenfunctions respectively. Let $\langle\ ,\ \rangle$ and $|\cdot|$ denote the inner product and the norm on H . Suppose we complete $\mathcal{C}_o^2$ in H with respect to the norm $\|\phi\|_1 = \{\sum_{n=1}^{\infty} \lambda_n \langle\phi,e_n\rangle^2\}^{1/2}$. The completion thus obtained is a (Sobolev) subspace H_o^1 of H , and its dual space $\mathcal{H}_o^{-1} = (\mathcal{H}_o^1)^*$ has the norm $\|\phi'\| = \{\sum_{n=1}^{\infty} \lambda_n^{-1} \langle\phi' , e_n\rangle^2\}^{1/2}$.

As to the Brownian motion (or Wiener process) , it seems natural to take

$$\dot{W}(t,x) = \sum_{n=1}^{\infty} \dot{b}_n(t)\, e_n(x) , \tag{2.5}$$

where $\{b_n(t)\}$ is a sequence of independent, identical copies of the standard Brownian motion in one dimension. However the process (2.5) cannot be realized in H , for, otherwise, the variance $E|W_t|^2$ would have been infinite .

Instead, since $\sum_{n=1}^{\infty} \frac{1}{\lambda_n} < \infty$, one may regard (2.5) as a H_o^{-1}-valued process for which $E\|W_t\|^2 = t\sum_{n=1}^{\infty} \frac{1}{\lambda_n}$ is defined. This is a naive way of conceiving the need for introducing the pair (H, H_o^{-1}) as a Gross' (abstract) Wiener space [1].

To recast the system (2·1) - (2·3) as an evolution equation, it is convenient to extend the domain of A_o . The extension $A:H_o^1 \to H_o^{-1}$ can be defined by the relation $(A\phi_1 , \phi_2) = \langle A^{1/2}\phi_1 , A^{1/2}\phi_2\rangle$ for $\phi_1 , \phi_2 \in H_o^1$, where $(\cdot , \cdot)$ stands for the duality between H_o^{-1} and H_o^1 . Then the system (2·1) - 2·3) may be regarded as an evolution equation in H_o^{-1}:

$$\frac{du}{dt} + Au = \dot{W}_t , \quad t > o$$
$$u(o) = \eta \in H . \tag{2.6}$$

From the above simple system, the exact solution can be found. In fact it is an Ornstein-Uhlenbeck process in H (see [1]). By the method of eigenfunction expansions, the system (2.6) is reducible to an infinite system of Itô equations in H_o^{-1}:

$$\frac{da_n}{dt} + \lambda_n a_n = \dot{b}_n(t) \ , \quad t > o \ , \qquad a_n(o) = c_n \ , \quad n = 1,2, \ldots \ , \tag{2.7}$$

where

$$a_n = \langle u \ , e_n \rangle \quad \text{and} \quad c_n = \langle \eta, e_n \rangle \ .$$

Now, suppose that one wishes to approximate the system (2.6) by a finite subsystem of (2.7) for $n \le N$. Then it is clear that this subsystem can be written as:

$$\frac{d}{dt} u_N + P_N A u_N = P_N \dot{W}_t \ , \quad t > o \ , \qquad u_N(o) = P_N \eta \ . \tag{2.8}$$

Here $u_N = P_N u$ and $P_N : H \to H_N = \text{sp.}\{e_1, \ldots, e_N\}$ is an orthogonal projection, and its extension to H_o^{-1} is also denoted by P_N for brevity.

The approximation (2.8) is a special consequence of Galenkin's method and its convergence in the mean to the solution of (2.6) can be proved easily by finding the error $\rho_N = u - u_N$. From (2.6) and (2.8) it is obvious that, with $P_N^{\perp} = I - P_N$,

$$\frac{d\rho_N}{dt} + P_N^{\perp} A \rho_N = P_N^{\perp} \dot{W}_t \ , \qquad t > o \ , \qquad \rho_N(o) = P_N^{\perp} \eta = \eta_N^{\perp} \ . \tag{2.9}$$

Similar to (2.6), the solution of (2.9) can be expressed as a stochastic integral

$$\rho_N(t) = e^{-tA_N^{\perp}} \eta_N^{\perp} + \int_o^t e^{-(t-s)A_N^{\perp}} dP_N^{\perp} W_s \tag{2.10}$$

If the initial state η is a H-valued random variable of finite variance, which is independent of $\{W_t\}$, the mean-square error can be computed to give

$$E|\rho_N(t)|^2 = E|e^{-tA_N^\perp}\eta_N^\perp|^2 + E|\int_o^t e^{-(t-s)A_N^\perp} dP_N^\perp W_s|^2$$
$$\leq e^{-2\lambda_N t} E|\eta_N^\perp|^2 + (\frac{1}{2\lambda_N})^2 \quad . \tag{2.11}$$

Thus, for $t > o$, the root-mean-square error is of $O(\frac{1}{\lambda_N})$.

The simple situation discussed above can be readily generalized as in the deterministic case in type of equation and methodology. Let us consider a generalized version of the system (2.6) in the Wiener space (H,V)

$$\frac{du}{dt} + Au = \Gamma\dot{W}_t \quad ,$$
$$u(o) = \eta \quad . \tag{2.12}$$

Let $V^* \subset H$ be the dual space of V, dense in H, and $A = V^* \to V$. $\Gamma \in \mathcal{L}^2(H)$ is a Hilbert-Schmidt operator. If A is a strictly self-adjoint operator whose eigenfunctions $\{e_n\}$ spans V^*, by applying Galerkin's method to (2.12), a parallel error analysis yields a slightly different estimate

$$E|\rho_N(t)|^2 \leq e^{-2\lambda_N t} E|\eta_N^\perp|^2 + (\frac{\mu_N}{2\lambda_N})^2 \quad , \tag{2.13}$$

where μ_N denotes the smallest-eigenvalue of $(P_N^\perp \Gamma)$ and λ_N is the same as before. Suppose that A generates a strongly continuous semigroup $\{T_t\}$ on H, but is not self-adjoint. The estimate (2.13) still holds if the spectrum $\sigma(A_N^\perp)$ lies in the right-half complex-plane with $\lambda_N = \inf.\ \mathrm{Re}\ \sigma(A_N^\perp)$ and $A_N^\perp = P_N^\perp A$.

As a generalization in another direction, let P_N, Q_N be two projectors on H onto H_N and K_N, respectively, which are assumed to be subspaces of V^*. Let us approximate the solution $u(t)$ of (2.12) by a random function $\tilde{u}(t)$ in H_N. The residual of this approximation is conventionally defined as

$$R_t(\tilde{u}) = \Gamma\dot{W}_t - (\frac{d\tilde{u}}{dt} + A\tilde{u}) \quad , \qquad t > o$$
$$R_o(\tilde{u}) = \eta - \tilde{u}(o) \quad . \tag{2.14}$$

In certain sense, one requires that the residual R_t be as small as possible. In the method of weighted residual, the "smallness" means the orthogonality condition:

$$Q_N R_t = o , \qquad \text{for some } Q_N . \tag{2.15}$$

This leads to, noting (2.14), the following system:

$$\begin{aligned} \frac{d}{dt} Q_N \tilde{u} + Q_N A \tilde{u} &= Q_N \Gamma \dot{W}_t , \qquad t > o \\ Q_N \tilde{u}(o) &= Q_N \eta \end{aligned} \tag{2.16}$$

Now, if we set $\tilde{u} = Q_N u$, the above system is just a Galerkin system. On the other hand, taking $\tilde{u} = P_N u$, the system (2.16) is the result of the so-called moment method (not to be confused with the moments in probability). In either case, an error estimate similar to (2.13) can be obtained.

In passing we remark that, although the approximation has been treated in a deterministic fashion, it can easily be rephrased in the probabilistic context. For example, consider the solution space to be $\mathcal{L}^2(C[o,T] \times H, p)$, the set of all H-valued random processes with continuous sample paths for $t \in [o,T]$, which have finite second moments w.r.t. the Wiener measure p. For an approximating function space, one may take e.g., $\mathcal{L}^2(C[o,T] \times H_N, p)$. Also it is possible to introduce methods of approximation based on a stochastic variational principle in an abstract Wiener space setting [3].

3. Approximate Moments for Bilinear Random Equations

A bilinear random equation is an abbreviation for the linear random equation with random coefficients. For example, in turbulence related problems [1], we have to consider the following bilinear random equation:

$$\frac{\partial u}{\partial t} - \alpha \Delta u = \dot{\xi}_t u , \tag{3.1}$$

and

$$\frac{\partial u}{\partial t} - \alpha \Delta u = \dot{\zeta}_t \cdot (\nabla u) , \tag{3.2}$$

where α is a constant, and ξ_t and ζ_t are scalar and vector random field. These equations are solved with a set of initial and boundary conditions. In the abstract setting indicated in Section 2, the equations can be put in the form:

$$\frac{du}{dt} + Au = B_t(u, \dot{\xi}_t), \qquad t \in (o,T)$$
$$u(o) = \eta \tag{3.3}$$

Again, the above equation is defined in the Wiener space (H,V) and $B:H \times U \to V$ is a bilinear operator, where U is a Banach space together with the Hilbert space K forming another Wiener space. In this Wiener space (K,U), the random process ξ_t is defined. To avoid the unnecessary complication, let us assume, throughout this section, $K = H$ and $U = V$.

It seems clear that, with an obvious modification, the method of eigenfunction expansions, or other methods of projections described in the previous section, can be used to construct an approximate solution to the system (3.3). However the actual computation of the approximate solution and the associated error estimate become much more difficult. On the other hand, it was shown that, if $\xi_t = W_t$ is a Brownian motion, all moments of solution to (3.1) or (3.2) can be determined by solving the related partial differential equations. In the remaining section, we shall do two things. First one will show by a simple technique that all moment equations can be derived for the general system (2.3) when $\xi_t = W_t$. Then the idea of approximating the equation (3.3) by the one with the coefficient ξ_t replaced by W_t.

In (3.3), setting $\xi_t = W_t$, we rewrite it in the integral form:

$$u_t = \eta - \int_o^t Au_s ds + \int_o^t \Gamma_s(u_s)dW_s, \tag{3.4}$$

where Γ_s is defined by the relation $\Gamma_s(u)v = B_s(u,v)$, $v \in V$. If $\Gamma_s(u) \in \mathcal{L}^2(H)$, $s \in (o,T)$, $u \in H$, the stochastic integral is well defined [1]. For a smooth function $\mathcal{J}$ on V, the Itô lemma [1] implies that

$$d\mathcal{J}(u_t) = \langle \Gamma_t^*(u_t)D\mathcal{J}(u_t), dW_t\rangle - (D\mathcal{J}(u_t), Au_t)dt$$
$$+ \frac{1}{2}\,\mathrm{tr}\,[\Gamma_t^* D^2\mathcal{J}(u_t)\,\Gamma_t]dt \tag{3.5}$$

where the symbol D means the H-differentiation, $(\ ,\ )$ the pairing of V and V^*, and tr. signifies the trace. By taking expectation of (3.5), it yields

$$\frac{dE\mathcal{T}(u_t)}{dt} + E(D\mathcal{T}(u_t)\ ,\ Au_t) = \frac{1}{2}\ \text{tr.}\ E\,[\Gamma_t^{\ *}D^2\mathcal{T}(u_t)\,\Gamma_t\,]\ , \tag{3.6}$$

in which $E\mathcal{T}$ means the expectation of $\mathcal{T}$.

As a special case, let $\mathcal{T}(u_t) = \langle h,u_t\rangle$, $h \in V^*$. The equation (3.6) yields an equation for the mean solution $M_t = Eu_t$:

$$\frac{dM_t}{dt} + AM_t = 0\ . \tag{3.7}$$

Next set $\mathcal{T}(u_t) = \langle h_1,u_t\rangle\,\langle h_2,u_t\rangle$ so that $E\mathcal{T} = \langle M_t^{(2)}h_1,h_2\rangle$ for $h_1,h_2 \in V^*$. Then (3.6) gives an equation for the correlation operator $M_t^{(2)}$, or the second moment of u_t :

$$\frac{dM_t^{(2)}}{dt} + (A \oplus A)M_t^{(2)} = (\Gamma_t^{\,*} \otimes \Gamma_t)M_t^{(2)}\ , \tag{3.8}$$

where $\oplus$ and $\otimes$ denote, respectively, the direct sum and the tensor product of two operators on appropriate tensor product spaces. In general, for $\mathcal{T}(u_t) = \prod_{j=1}^{n} \langle h_j,u_t\rangle$, the n-th moment $M_t^{(n)}$ is defined by $M_t^{(n)}(h) = E\prod_{j=1}^{n} \langle h_j,u_t\rangle$ as a linear form on n-tensor product of H. It is easily shown that $M_t^{(n)}$ satisfies an n-fold version of (3.8). One must note, however, that these moment equations hold in the weak sense.

Now suppose that the coefficient ξ_t is a non-anticipating functional of the Browniam motion W_t with $E\xi_t = 0$ and $E|\xi_t|^2 < \infty$, and it is representable as a stochastic integral

$$\xi_t = \int_0^t \Phi_s(\omega)dW_s\ , \tag{3.9}$$

where $\Phi_s(\omega)$ with $E\,\Phi_s = 0$ is a non-anticipating Hilbert-Schmidt operator-valued function of $\{W_t\}$. Rewrite the system (3.3) as

$$u_t = \eta - \int_0^t Au_s ds + \int_0^t \Gamma_s(u_s)d\xi_s \tag{3.10}$$

and consider the approximation

$$\tilde{u}_t = \eta - \int_o^t A\tilde{u}_s ds + \int_o^t \tilde{\Gamma}_s(\tilde{u}_s)dW_s . \tag{3.11}$$

Here Γ_s is defined as before and $\tilde{\Gamma}_s$ is an approximating operator to be found. It will be so determined as to give a "small" residual for this approximation

$$R_t(\tilde{u}) = \int_o^t \Gamma_s(\tilde{u}_s)d\xi_s - \int_o^t \tilde{\Gamma}_s(\tilde{u}_s)dW_s . \tag{3.12}$$

Since $ER_t = o$, by "smallness" here we mean that

$$E|R_t(v)|^2 = o , \quad t \in [o,T] , \tag{3.13}$$

for every $v_s \in \mathcal{F}_s^1$, the σ-field generated by $\{W_\tau, \tau > s\}$. Then, by taking (3.9) and (3.12) into account, the condition (3.13) implies that

$$\text{tr.} \{ E[\Phi_s{*}\Gamma_s{*}(\cdot)\Gamma_s(\cdot)\ \Phi_s] - \tilde{\Gamma}{*}(\cdot)\tilde{\Gamma}(\cdot)] \ (Ev_s \otimes v_s) = 0 \tag{3.14}$$

or we may choose

$$\tilde{\Gamma}_s = \{E(\Phi_s{*}\ \Gamma_s{*}\Gamma_s\ \Phi_s)\}^{1/2} \tag{3.15}$$

It is not difficult to get a mean-square error estimate from (3.10) and (3.11), depending on the sizes of the approximations $\tilde{u}_t$ and $\tilde{\Gamma}_t$.

Finally, if ξ_t is a Wiener functional, but not representable as the stochastic integral (3.9), the approximations by the stochastic integral equation can still be made. However, in this case, a rational choice of $\tilde{\Gamma}_t$ such as (3.15) may not be possible. Instead one must choose $\tilde{\Gamma}_t$ judiciously by an educated guess.

4. Stochastic Modeling and Other Methods

From a mathematical viewpoint, it is necessary to show the approximate solution is close to the true solution in a certain sense. In application, such a proof is often not possible. Even if this goal is achievable, the method may be of little practical value. An outstanding example is the simple-minded iteration or perturbation procedure. Within the limit of computational feasibility the first few terms of approximation are meaningless due to the slow rate and, sometimes, the nonuniformity of convergence. To cope with this kind of difficulty, various formal methods have been introduced. Most of them are under the name of uniform asymptotic or modified perturbation method [4]. Another method of solving the bilinear and nonlinear evolution equation of Gaussian coefficients is the homogeneous chaos or Wiener-Hermite expension [5]. These methods will not be described here. Instead we shall give a brief discussion of a novel method known as the direct interaction or random coupling model. This approximation was shown to be valid for a random bilinear equation [6]. In general the nature of approximation is not known. In an attempt to give it a mathematical basis, we envisioned it as a statistical estimation scheme. However we have not yet been able to develop a logically consistent theory. Some of our basic ideas will be indicated as follows.

To be specific, consider the nonlinear random evolution equation in (H,V):

$$\begin{aligned} &\frac{du}{dt} + Au + B(u,u) = \Gamma \dot{W}_t , \qquad t > o \\ &u(o) = \eta , \end{aligned} \tag{4.1}$$

where $B: V^* \times V^* \to V$ is a bilinear operator. In particular the system may represent the Langevin-Navier-Stokes equations in turbulence [1]:

$$\frac{\partial u}{\partial t} + K(u \cdot \nabla)u = \upsilon \Delta u + \Gamma \dot{W}_t , \tag{4.2}$$

where K is a linear integral operator.

To estimate certain parameters of the solution distribution, let us imagine that a random sample $S_n = \{u_1, u_2, \ldots , u_n\}$ of size n is taken from the solution population in H in the presence of noises. Then the parameters are estimated based on suitable statistics of the sample. In the process, we must model the "black box", or the system $\mathcal{S}$ that generates the noisy data.

Suppose that we wish to estimate the first two moments $M_1 = Eu$, $M_2 = E(u \otimes u)$. Denote their estimators by $\hat{M}_1$, $\hat{M}_2$ respectively. In consistence with the equation (4.1), they are required to satisfy the equation:

$$\begin{aligned} &\frac{d\hat{M}_1}{dt} + A\hat{M}_1 + B\hat{M}_2 = o , \qquad t > o \\ &\hat{M}_1(o) = E\eta . \end{aligned} \tag{4.3}$$

For the estimator $\hat{M}_2$, we let

$$\hat{M}_2 = \varphi(u_1, u_2, \ldots, u_n) \tag{4.4}$$

where φ is a statistic and, as before, no distinction has been made between the random functions u_j , and their sample functions . Now, in view of (4.4), it seems natural to model the system $\mathscr{S}$ as

$$\begin{aligned} &\frac{du_j}{dt} + Au_j + [B\,\varphi(u_1, \ldots, u_n)]_j = \Gamma_j \dot{W}_t , \\ &u_j(o) = \eta , \qquad j = 1,2, \ldots , n , \end{aligned} \tag{4.5}$$

in which $\Gamma_j \dot{W}_t$ are independent identical Gaussian white noises. Within this model, we shall call the estimator $\hat{M}_2$ "unbiased" if

$$E\hat{M}_2 = E(u_1 \otimes u_1) = \ldots = E(u_n \otimes u_n) . \tag{4.6}$$

It is "consistent" if

$$E\hat{M}_2 = \lim_{n\to\infty} \varphi(u_1, \ldots, u_n) \quad \text{a.s.} \tag{4.7}$$

For example, if we choose the statistic

$$\varphi = \frac{1}{n} \sum_{j=1}^{n} \varphi_j , \tag{4.8}$$

and

$$\varphi_j(u_1, \ldots, u_n) = E(u_j \otimes u_j) + \frac{1}{n}\sum_{k,l=1}^{n} \dot{b}_{jkl}(t)\,(u_k \otimes u_l) \tag{4.9}$$

If $b_{jkl}(t)$ are independent Brownian motions in one-dimension, symmetric with respect to the indices, then the estimator satisfies the properties (4.6) and (4.7) . For this estimator, the system (4.5) with $[B\,\varphi]_j = B\,\varphi_j$ yields the random coupling model presented in [1]. The remarkable thing about this model is that, as $n \to \infty$, the joint distribution of u_j can be found. The limit distribution is Gaussian with the mean function determined by the equation (4.3) and the second moment $\hat{M}_2$ satisfying a known equation.

The statistical theory outlined above is unsatisfactory, because there are gaps. The most serious ones are the definitions of unbiased and consistent estimates (4.6) and (4.7) which are based on the model $\mathscr{S}$. Also it seems rather difficult to choose a "best" estimator, such as a minimum variance, for the nonlinear problem, such that the resulting model is solvable.

REFERENCES

1. Chow, P.L. (1978) Stochastic Partial Differential Equations in Turbulence Related Problems, in Probabilistic Analysis and Related Topics, ed. by A.T. Bharuchu-Reid, Academic Press, New York.

2. Dawson, D.A. (1975) Stochastic Evolution Equations and Related Measure Processes, J. Multivari. Analy. 5, 1-52.

3. Kuo, H.H. (1973) On the Stochastic Maximum Principle in Banach Space, J. Functional Analy. 14, 146-161.

4. Chow, P.L. (1975) Perturbation Methods in Stochastic Wave Propagation, SIAM Review, 17, 57-81.

5. Wiener, N. (1957) Nonlinear Problems in Random Theory, MIT Press, Cambridge, Mass.

6. Chow, P.L. (1974) On the Exact and Approximate Solutions of a Random Parabolic Equation, SIAM J. Appl. Math., 27, 376-397.

A Method of Averaging for Random Differential Equations with Applications to Stability and Stochastic Approximations

Stuart Geman

Division of Applied Mathematics
Brown University, Providence, Rhode Island 02912

1. Introduction
2. Preliminary Remarks on Mixing Processes
3. The Basic Lemma
4. Approximating the Solutions of Some Random Differential Equations
5. Stability in Random Homogeneous Linear Systems
6. Continuous Time Stochastic Approximation

References

1. Introduction. Classically, the "method of averaging" treats the following problem. Suppose $H: R^n \times R^1 \to R^n$ such that, for some G,

$$\lim_{T\to\infty} \frac{1}{T} \int_0^T H(x,t)\,dt = G(x) \tag{1}$$

for all x. Consider the differential equations

$$\dot{x}_\varepsilon(t) = \varepsilon H(x_\varepsilon(t),t) \qquad x_\varepsilon(0) = x_0, \tag{2}$$

and

$$\dot{y}_\varepsilon(t) = \varepsilon G(y_\varepsilon(t)) \qquad y_\varepsilon(0) = x_0. \tag{3}$$

Under what conditions will $y_\varepsilon(t)$ uniformly approximate $x_\varepsilon(t)$ as $\varepsilon \to 0$; either on finite intervals (with length of order $1/\varepsilon$) or on the infinite interval $t \geq 0$?* In this paper, we will discuss some stochastic generalizations.

There is a natural analogue to the deterministic averaging problem within the context of random differential equations. Suppose that for each x and t the right-hand side of (2) is a random, rather than deterministic, vector: $H = H(x,\omega,t)$, where ω is a sample point in a probability space. For the "averaged equation", (3), we use the expected value of H:

$$G(x) = E[H(x,\omega,\tau)], \qquad \text{for each } x \text{ and } t \text{ fixed},$$

assuming for now that this is independent of t. (1) suggests ergodicity, and so we might assume that H is ergodic (say, for each fixed x) and ask the same question: when does the (deterministic) function $y_\varepsilon(t)$ approximate the (random) function $x_\varepsilon(t,\omega)$, as ε gets small? Actually, for our purposes ergodicity is not the right property. We need a stronger form of mixing (namely "strong mixing") but we do not need stationarity. As discussed in section 2, many of the Markov processes, for example, are strong mixing.

Section 2 presents a rigorous discussion of the strong mixing property (really, "properties", since strong mixing has several nonequivalent formulations). But, for the purposes of this introduction, we will say, loosely, that a strong mixing process is one for which the "past" and "future" are asymptotically independent. So, in the above example, we would assume that $H(x,\omega,t)$ is nearly independent of $H(x,\omega,s)$ whenever

---

*Mitropolsky [21] is recommended for a thorough history of the averaging method, as applied to deterministic equations.

$|t-s|$ is large. In specific applications, $H(x,\omega,t)$ is often of the form $J(x,\phi(t,\omega),t)$, where ϕ is some possibly vector-valued process. Here, the condition is that ϕ itself be strongly mixing.

Now let us look more closely at a random version of (2):

$$\dot{x}_\varepsilon(t,\omega) = \varepsilon H(x_\varepsilon(t,\omega),\omega,t) \qquad x_\varepsilon(0,\omega) = x_0 \in R^n \tag{4}$$

where H is strongly mixing in t. If we make the change of variables $t \to t/\varepsilon$, then (4) becomes

$$\dot{x}_\varepsilon(t,\omega) = H(x_\varepsilon(t,\omega),\omega,t/\varepsilon). \tag{5}$$

As $\varepsilon \to 0$, the "mixing rate" of $H(x,\omega,t/\varepsilon)$ increases, such that for a fixed separation in time H is more and more nearly independent of itself. Consequently, H fluctuates more and more rapidly as $\varepsilon \to 0$, and we might expect that $X_\varepsilon(t,\omega)$, the integral of H, will come to reflect only the average of these fluctuations. If we "average" (5) (take expected value, treating $x_\varepsilon(t,\omega)$ as constant) we get

$$\dot{y}(t) = G(y(t)), \qquad y(0) = x_0. \tag{6}$$

On finite intervals, $[0,T]$, the "approximation" (to be made precise below) of $x_\varepsilon(t,\omega)$ by $y(t)$ holds quite generally, and for many equations the interval can be extended to $[0,\infty)$. Of course, $[0,T]$ in (5) corresponds to $[0,T/\varepsilon]$ in (4), the usual interval of approximation achived in deterministic averaging.

The problem of approximating the solution of a random mixing equation by the solution of its associated averaged equation is the topic of section 4. We will generalize slightly, allowing for a slow, as well as fast, time scale, and allowing the averaged equation to be nonautonomous. Specifically, we will discuss the equations

$$\dot{x}_\varepsilon(t,\omega) = F(x_\varepsilon(t,\omega),\omega,t,t/\varepsilon) \qquad x_\varepsilon(0,\omega) = x_0,$$

and

$$\dot{y}_\varepsilon(t) = G_\varepsilon(y_\varepsilon(t),t) \qquad y_\varepsilon(0) = x_0,$$

where

$(\Omega,\mathscr{F},P)$ is a probability space,

$F\colon R^n \times \Omega \times R^1 \times R^1 \to R^n$, $F(x,\omega,t,\tau)$ is strongly mixing in τ, and

$G_\varepsilon(x,t) = E[F(x,\omega,t,t/\varepsilon)]$.

Quite generally, the following form of averaging is in force: for every $T > 0$

$$\sup_{t\in[0,T]} |x_\varepsilon(t) - y_\varepsilon(t)| \to 0 \quad \text{in probability,} \tag{7}$$

as $\varepsilon \to 0$ (Theorem 2). But the main purpose of section 4 (Theorems 1 and 3) is to establish conditions for

$$\lim_{\varepsilon\to 0} \sup_{t\geq 0} E|x_\varepsilon(t) - y_\varepsilon(t)|^2 = 0. \tag{8}$$

Section 5 discusses stability in the linear homogeneous system

$$\dot{x}(t,\omega) = A(t,\omega)x(t,\omega) \tag{9}$$

where A is an $n \times n$ matrix-valued random process. If $B(t) = E[A(t)]$, then the averaged equation is

$$\dot{y} = B(t)y. \tag{10}$$

For suitable A, the asymptotic stability of (10) is sufficient for the almost sure and L^2 asymptotic stability of (9) (Theorem 4).

Since the results of sections 4 and 5 will appear elsewhere (in [9]), their proofs are omitted.

In section 6, averaging techniques are applied to some problems of stochastic approximation. Stochastic approximation procedures are recursive algorithms for the estimation of parameters from a sequence of random observations. We will look at two such procedures which have previously been formulated in discrete time, and analyze their continuous time analogues. Here again a type of averaging is at work, and by exploiting it (in Theorems 5 and 6) we can prove consistency when the observations (now a continuous time process) are strongly mixing. This extends the application of these procedures to, for example, a variety of Markov processes. A number of recent publications have been concerned with generalizing the application of stochastic approximation procedures. Their relation to the present results will be discussed in section 6. (Sections 4 and 5 also include discussions of relevant literature.)

What really ties all of this together is that each of these results is an application of the same lemma; namely, Lemma 1, which is the topic of section 3. In a very general context, this lemma relates the solution of a random equation to that of its averaged equation, whether or not mixing processes are involved. The proof of each theorem in this paper utilizes Lemma 1 to evaluate the error in approximating the random solution by the deterministic averaged solution. Although it will appear elsewhere (in [9]), the proof of the lemma is short and simple, and is repeated here.

2. Preliminary Remarks on Mixing Processes. This section serves as a brief introduction to strong mixing. We will begin with notation and definitions, and then discuss some consequences of the mixing property which will be needed later. Also included are a few examples of strongly mixing processes.

In this paper, reference will be made to three distinct types of mixing. Their definitions are our first order of business.

Given the probability space $(\Omega, \mathcal{F}, P)$, let $\mathcal{F}_0^t \subset \mathcal{F}$ and $\mathcal{F}_t^\infty \subset \mathcal{F}$ be two families of σ-fields indexed by $t \geq 0$. Mixing will be defined in terms of these families, and, in each application, these families will be defined in terms of the right-hand side of the stochastic system. (For now, think of $\mathcal{F}_a^b$ as the sigma field generated by some stochastic process between the times a and b, any $a \leq b$.)

In formulating and utilizing the various strong mixing conditions, it is convenient to introduce a certain signed measure: for every $t \geq 0$ and $\delta \geq 0$, let $v_{t,\delta}(\cdot)$ be the measure, on $(\Omega \times \Omega, \mathcal{F}_0^t \times \mathcal{F}_{t+\delta}^\infty)$, defined by

$$v_{t,\delta}(B) = P(\omega\colon (\omega,\omega) \in B) - P \times P(B)$$

$B \in \mathcal{F}_0^t \times \mathcal{F}_{t+\delta}^\infty$. Since $\{B \in \mathcal{F}_0^t \times \mathcal{F}_{t+\delta}^\infty\colon (\omega\colon (\omega,\omega) \in B) \in \mathcal{F}\}$ is a monotone class which contains the elementary sets, $(\omega\colon (\omega,\omega) \in B)$ is in $\mathcal{F}$ for all $B \in \mathcal{F}_0^t \times \mathcal{F}_{t+\delta}^\infty$, so the definition makes sense. Notice that $v_{t,\delta} \equiv 0$ whenever $\mathcal{F}_0^t$ and $\mathcal{F}_{t+\delta}^\infty$ are independent with respect to P (extend from the rectangles). Loosely speaking, the total variation of $v_{t,\delta}$ will be small whenever $\mathcal{F}_0^t$ and $\mathcal{F}_{t+\delta}^\infty$ are "nearly" independent.

Here are three versions of "strong mixing":

Type I. $\rho(\delta) \equiv \sup_{t \geq 0} \sup_{\substack{A \in \mathscr{F}_0^t \\ B \in \mathscr{F}_{t+\delta}^\infty \\ P(A) > 0}} \frac{1}{P(A)} |v_{t,\delta}(A \times B)| \to 0$

as $\delta \to \infty$.

Type II. $\rho(\delta) \equiv \sup_{t \geq 0} \sup_{A \in \mathscr{F}_0^t \times \mathscr{F}_{t+\delta}^\infty} |v_{t,\delta}(A)| \to 0$

as $\delta \to \infty$.

Type III. $\rho(\delta) \equiv \sup_{t \geq 0} \sup_{\substack{A \in \mathscr{F}_0^t \\ B \in \mathscr{F}_{t+\delta}^\infty}} |v_{t,\delta}(A \times B)| \to 0$

as $\delta \to \infty$.*

The first two were introduced by Kolmogorov (see Volkonskii and Rozanov [28]), and the last by Rosenblatt [23]. Type I has been frequently used in the context of stochastic differential equations (cf. Khasminskii [13], Cogburn and Hersh [5], and Papanicolaou and Kohler [22]). It is not hard to show that I $\Longrightarrow$ II $\Longrightarrow$ III, and further implications can be disproved by counterexample. If $\{\mathscr{F}_a^b\}$ is generated by a stationary process, then any of the strong mixing conditions implies ergodicity.

$|v|_{t,\delta}(\cdot)$ will refer to the total variation measure. Since $v_{t,\delta}$ is the difference of two probability measures:

$|v|_{t,\delta}(\Omega \times \Omega) \leq 2$, and

$$|v|_{t,\delta}(\Omega \times \Omega) = 2 \sup_{A \in \mathscr{F}_0^t \times \mathscr{F}_{t+\delta}^\infty} |v_{t,\delta}(A)|.$$

So, with Type II (or I) mixing:

$$|v|_{t,\delta}(\Omega \times \Omega) \leq 2\rho(\delta).$$

*Processes which generate σ-fields satisfying this condition are sometimes called "completely regular".

Consequently, if $f(\omega,\eta)$ is $\mathscr{F}_0^t \times \mathscr{F}_{t+\delta}^{\infty}$ measurable and $|f(\omega,\eta)| \le c$ for all $(\omega,\eta) \in \Omega \times \Omega$, then

$$\left|\int_{\Omega\times\Omega} f(\omega,\eta)\,dv_{t,0}\right| = \left|\int_{\Omega\times\Omega} f(\omega,\eta)\,dv_{t,\delta}\right| \tag{11}$$

$$\le \int_{\Omega\times\Omega} |f(\omega,\eta)|\,d|v|_{t,\delta} \le 2c\rho(\delta).$$

If we specialize to Type I mixing then (see Billingsley [1], Chapter 4):

$$\left|\int_{\Omega\times\Omega} f(\omega)g(\eta)\,dv_{t,\delta}\right| \le 2c\rho(\delta)E|f| \tag{12}$$

whenever f and g are $\mathscr{F}_0^t$ and $\mathscr{F}_{t+\delta}^{\infty}$ measurable respectively, $E|f| < \infty$, and $|g| \le c$ a.s.

Analogous definitions are used for discrete parameter processes $(x_0, x_1, \ldots)$, with $\mathscr{F}_n^m$ (the σ-field generated by $\{x_i\colon\ n \le i \le m\}$) replacing $\mathscr{F}_a^b$. If $\{x_n\}$ satisfies one of the strong mixing conditions, then, obviously, so does the continuous parameter process

$$x(t) = x_n \quad \text{when} \quad n \le t < n + 1. \tag{13}$$

It is also obvious that functions of mixing processes, of the form $y(t) = f(x(t))$, are themselves mixing, and of the same type. Of course, a sequence of independent random variables is mixing in every sense, as is an "m-dependent" sequence - i.e., one for which $\mathscr{F}_0^n$ is independent of $\mathscr{F}_{n+k}^{\infty}$ whenever $k > m$. Similarly, if $B(t)$ is a Brownian motion, then $y(t) = B(t+\tau) - B(t)$ is strongly mixing by all three definitions.

For Type III mixing, nontrivial examples include many of the stationary Gaussian processes (discrete and continuous parameter). Sufficient conditions for a Gaussian process to be Type III mixing, given in terms of the spectral density, are in Rozanov ([24], Chapter 4, section 10). For Type I mixing, Markov processes provide a wide variety of examples. Specifically, if $x(t)$ is an ergodic Markov process satisfying Doeblin's condition, then $x(t)$ is Type I mixing, and, in fact,

$$\rho(t) \le \gamma e^{-\lambda t} \tag{14}$$

for some γ and $\lambda > 0$. And, similarly, if $\{x_n\}$ is ergodic Markov, satisfies Doeblin's condition, and is aperiodic, then $\{x_n\}$ is Type I mixing and

$$\rho(n) \leq \gamma e^{-\lambda n}.$$

These last two assertions follow easily from Doob's ([6], Chapters V and VI) treatment of Markov processes. Any finite state Markov process, for example, satisfies Doeblin's condition, but the condition is much broader than this. See Doob ([6], especially pp. 192-194) for a thorough discussion of Doeblin's condition.

In addition to a mixing assumption, we will sometimes impose a condition on the "mixing rate", in the form of

$$\sum_{n=1}^{\infty} \rho(e^{\varepsilon n}) < \infty \qquad \forall \varepsilon > 0. \tag{15}$$

In view of (14), (15) certainly holds for the Markov processes discussed above.

3. The Basic Lemma. Our purpose here is to derive an expression which relates in a convenient manner the solution of a random differential equation to that of its averaged equation. This expression (Lemma 1, below) plays a central role in obtaining each of our results. After its proof, we will outline, heuristically, how it is applied to the averaging problem.

For convenience, we will generally not distinguish between "for almost every ω" and "for every ω". Look at the systems

$$\dot{x}(t,\omega) = H(x(t,\omega),\omega,t) \qquad x(0,\omega) = x_0 \in R^n$$

$$\dot{y}(t) = G(y(t),t) \qquad y(0) = x_0$$

where

(1) H is jointly measurable in its three arguments.

(2) $G(x,t) = E[H(x,\omega,t)]$, and for all i and j $\frac{\partial}{\partial x_j} G_i(x,t)$ exists, and $G(x,t)$ and $\frac{\partial}{\partial x_j} G_i(x,t)$ are continuous (in (x,t)).

(3) For some $T > 0$:

(a) For every ω, there exists a unique absolutely continuous solution, $x(t,\omega)$, on $[0,T]$.

(b) A solution to
$$\frac{\partial}{\partial t} g(t,s,x) = G(g(t,s,x),t), \quad g(s,s,x) = x,$$
exists on $[0,T] \times [0,T] \times R^n$.

We will use the following notation:

(1) $g_s(t,s,x) = \frac{\partial}{\partial s} g(t,s,x)$.

(2) $g_x(t,s,x) =$ the $n \times n$ matrix with (i,j) component $\frac{\partial}{\partial x_j} g_i(t,s,x)$.

(3) If $K: R^n \to R^1$ and $K \in C^1$,

$$K'(x) = (\frac{\partial}{\partial x_1} K(x),\ldots,\frac{\partial}{\partial x_n} K(x)),$$

and

$$\frac{\partial}{\partial x} K(g(t,s,x(s,\omega))) = K'(g(t,s,x(s,\omega)))g_x(t,s,x(s,\omega)).$$

(4) Define the families of σ-fields $\mathscr{F}_0^t$ and $\mathscr{F}_t^\infty$ such that, for each $t \geq 0$, $\mathscr{F}_0^t$ contains the σ-field generated by

$$\{H(x,\omega,s): 0 \leq s \leq t, -\infty < x < \infty\},$$

and $\mathscr{F}_t^\infty$ contains the σ-field generated by

$$\{H(x,\omega,s): t \leq s < \infty, -\infty < x < \infty\}.$$

Two facts that will be needed are:

(1) $g_s(t,s,x) = -g_x(t,s,x)G(x,s)$ for all $t \in [0,T)$, $s \in [0,T)$, and $x \in R^n$ (cf. Hartman [10], Chapter 5), and

(2) For any $f: \Omega \times \Omega \to R^1, \mathscr{F}_0^t \times \mathscr{F}_t^\infty$ measurable,

$$\int_{\Omega\times\Omega} \{f(\omega,\omega) - f(\omega,\eta)\}dP(\omega)dP(\eta) = \int_{\Omega\times\Omega} f(\omega,\eta)dv_{t,0}$$

(follows easily from the definition of $v_{t,0}$, and a monotone class argument for functions on $\Omega \times \Omega$).

<u>Lemma 1</u>. <u>For any</u> C^1 <u>function</u> $K: R^n \to R^1$ <u>and</u> $t \in [0,T)$:

$$E[K(x(t))] = K(y(t)) + \int_0^t \int_{\Omega\times\Omega} (\frac{\partial}{\partial x} K(g(t,s,x(s,\omega)))) \cdot H(x(s,\omega),\eta,s)dv_{s,0}ds, \tag{16}$$

<u>provided that</u>

$$\left(\frac{\partial}{\partial x} K(g(t,s,x(s,\omega)))\right)\cdot H(x(s,\omega),\eta,s),$$

<u>and</u>

$$\left(\frac{\partial}{\partial x} K(g(t,s,x(s,\omega)))\right)\cdot H(x(s,\omega),\omega,s)$$

<u>are absolutely integrable on</u> $\Omega \times \Omega \times [0,T]$, <u>with respect to</u> $dP(\omega)dP(\eta)ds$.

<u>Proof</u>.

$$K(x(t,\omega)) - K(y(t)) = \int_0^t \frac{\partial}{\partial s} K(g(t,s,x(s,\omega)))ds$$

$$= \int_0^t K'(g(t,s,x(s,\omega)))\cdot\{g_x(t,s,x(s,\omega))\dot{x}(s,\omega) + g_s(t,s,x(s,\omega))\}ds$$

$$= \int_0^t K'(g(t,s,x(s,\omega)))\cdot\{g_x(t,s,x(s,\omega))H(x(s,\omega),\omega,s)$$

$$- g_x(t,s,x(s,\omega))G(x(s,\omega),s)\}ds$$

$$= \int_0^t \int_\Omega \left(\frac{\partial}{\partial x} K(g(t,s,x(s,\omega)))\right)\cdot\{H(x(s,\omega),\omega,s) - H(x(s,\omega),\eta,s)\}dP(\eta)ds.$$

Hence,

$$E[K(x(t))] = K(y(t)) + \int_0^t \int_{\Omega\times\Omega} \left(\frac{\partial}{\partial x} K(g(t,s,x(s,\omega)))\right)\cdot$$

$$\{H(x(s,\omega),\omega,s) - H(x(s,\omega),\eta,s)\}dP(\eta)dP(\omega)ds$$

$$= K(y(t)) + \int_0^t \int_{\Omega\times\Omega} \left(\frac{\partial}{\partial x} K(g(t,s,x(s,\omega)))\right)\cdot H(x(s,\omega),\eta,s)dv_{s,0}ds. \quad \text{Q.E.D.}$$

<u>Remarks</u>:

(1) Various generalizations will be used. The proof is the same for any of the following:

(a) The initial conditions can be random $(x_0 \to x_0(\omega))$, if we assume $K(y(t)) \in L^1$ and write $E[K(y(t))]$ in place of $K(y(t))$ in (16).

(b) The function K can be random: $K\colon R^n \times \Omega \to R^1$, provided that $K(x,\omega)$ is C^1 for every $\omega \in \Omega$. Here again, we assume $K(y(t)) \in L^1$ and replace $K(y(t))$ by $E[K(y(t))]$.

(c) There is no difficulty in including processes whose paths have isolated jumps, provided that G and $\frac{\partial}{\partial x_j} G_i$ have discontinuities, at most, at isolated t's. (Of course, the equations for $\dot{x}(t,\omega)$ and $\frac{\partial}{\partial t} g(t,s,x)$ will not, in general, be satisfied at these discontinuities.) So, for example, the results in this paper apply to differential equations involving discrete time processes which have been interpolated into continuous time processes, as in (13).

(2) For the stochastic differential equation

$$dx = b(x,t)dt + A(x,t)dW \qquad x(0) = x_0 \in R^n,$$

(16) has a natural analogue:

$$E[K(x(t))] = K(y(t)) + \frac{1}{2}\int_0^t E[\mathrm{tr}\{A^T(x(s),s)\frac{\partial^2}{\partial x^2} K(g(t,s,x(s)))A(x(s),s)\}]ds$$

where we have defined $\dot{y} = b(y,t)$ as the "averaged equation". The proof is the same, except that Itô's formula is used to evaluate $dK(g(t,s,x(s)))$.

Here is an example of how the lemma is used. Return to equations (5) and (6), and let $\mathscr{F}_a^b$ be the σ-field generated by

$$\{H(x,\omega,t): \ a \leq t \leq b, \ -\infty < x < \infty\},$$

using strict inequality for a or b infinite. Define $v_{t,\delta}$ as in section 2, and assume Type II mixing. Letting $K(x) = |x-y(t)|^2$ in Lemma 1:

$$E|x_\varepsilon(t) - y(t)|^2 = \int_0^t \int_{\Omega\times\Omega} (\frac{\partial}{\partial x} K(g(t,s,x_\varepsilon(s,\omega)))) \cdot H(x_\varepsilon(s,\omega),\eta,s/\varepsilon)dv_{s/\varepsilon,0}ds.$$

Then, when we can, we write (for any small $\delta > 0$)

$$E|x_\varepsilon(t) - y(t)|^2 = O(\delta) + \int_\delta^t \int_{\Omega\times\Omega} (\frac{\partial}{\partial x} K(g(t,s,x_\varepsilon(s-\delta,\omega)))) \cdot H(x_\varepsilon(s-\sigma,\omega),\eta,s/\varepsilon)dv_{s/\varepsilon,0}ds = O(\delta)$$

$$+ \int_\delta^t \int_{\Omega\times\Omega} (\frac{\partial}{\partial x} K(g(t,s,x_\varepsilon(s-\delta,\omega)))) \cdot H(x_\varepsilon(s-\sigma,\omega),\eta,s/\varepsilon)dv_{(s-\delta)/\varepsilon,\delta/\varepsilon}ds$$

(since $x_\varepsilon(s-\delta)$ is $\mathscr{F}_0^{(s-\delta)/\varepsilon}$ measurable)

$$= O(\delta) + O(\rho(\delta/\varepsilon)).$$

Finally, then,

$$\lim_{\varepsilon\to 0} E|x_\varepsilon(t) - y(t)|^2 = O(\delta)$$

$$\Rightarrow \lim_{\varepsilon\to 0} E|x_\varepsilon(t) - y(t)|^2 = 0,$$

and, hopefully, uniformly on some t interval.

4. Approximating the Solutions of some Random Differential Equations.

Stochastic versions of averaging fall broadly into two categories. On the one hand are those averaging techniques developed for Itô-type equations, whereas on the other, reference has been to stochastic equations involving a smoother mixing, such as we are concerned with here. Results of the first type are not particularly relevant, so our discussion in this regard will be brief.

Lybrand [20] studies systems of the form

$$dx_\varepsilon = \varepsilon[Ax_\varepsilon dt + F(x_\varepsilon,t)dB] \qquad x_\varepsilon(0) = x_0$$

with A a constant or periodic matrix, and B a vector Brownian motion. Here the averaged equation is

$$dy_\varepsilon = \varepsilon Ay_\varepsilon dt \qquad y_\varepsilon(0) = x_0,$$

so that $y_\varepsilon(t)$ is the mean of the random solution. Under suitable growth conditions on F, $y_\varepsilon(t)$ will uniformly approximate $x_\varepsilon(t)$ (in the sense of (7)) on intervals with length of order $1/\varepsilon^h$ (any $0 < h < 1$), as $\varepsilon \to 0$. If A, or its time average, is a stability matrix, then this approximation extends to $[0,\infty)$. Vrkoc [29] treats

$$dx_\varepsilon = \varepsilon a(t,x_\varepsilon)dt + \phi(\varepsilon)B(t,x_\varepsilon)dw_\varepsilon$$

where

(a) $w_\varepsilon(t)$ can be any of a class of vector-valued independent increment processes (including Brownian motion), possibly depending on ε,

(b) the function $\phi(\varepsilon)$ depends on the particular $w_\varepsilon(t)$ ($\phi(\varepsilon) = \sqrt{\varepsilon}$ when $w_\varepsilon(t)$ = Brownian motion)

(c) a and B have appropriate time averages in a sense similar to (1).

In Vrkoc's work, the averaged equation is itself stochastic with a and B replaced by their time averages, and w_ε replaced by an associated stochastic process. One conclusion is that

$$\sup_{t\in[0,T/\varepsilon]} |x_\varepsilon(t) - y(\varepsilon t)| \to 0 \quad \text{in} \quad L^2$$

where $y(t)$ solves the averaged equation. And, when the averaged equation possesses certain stability properties, there are also $[0,\infty)$ results. (See the articles by these authors for some further references on averaging in Itô-type equations.)

Two 1966 papers by Khasminskii ([12], [13]) treat the problem of averaging in random mixing equations. In the first [12], Khasminskii examines systems of the form of (4):

$$\dot{x}_\varepsilon = \varepsilon H(x_\varepsilon, \omega, t) \qquad x_\varepsilon(0) = x_0,$$

beginning with an ergodic-type hypothesis reminiscent of the classical assumption (1):

$$\lim_{T\to\infty} \sup_{t>0} E[|\frac{1}{T}\int_t^{t+T} H(x,\omega,s)ds - G(x)|] = 0$$

for all x and some function G. If $y_\varepsilon(t)$ solves the averaged equation

$$\dot{y}_\varepsilon = \varepsilon G(y_\varepsilon) \qquad y_\varepsilon(0) = x_0,$$

then, with some boundedness and smoothness conditions on H, the conclusion is

$$\lim_{\varepsilon\to 0} \sup_{t\in[0,T/\varepsilon]} E[|x_\varepsilon(t) - y_\varepsilon(t)|] = 0$$

for all $T > 0$. The main purpose of the article, however, is an investigation of the asymptotic (small ε) distribution of

$$\frac{x_\varepsilon(t) - y_\varepsilon(t)}{\sqrt{\varepsilon}}$$

on intervals $[0,T/\varepsilon]$. With appropriate assumptions, including a mixing condition for H, the normalized error, $(x_\varepsilon(t)-y_\varepsilon(t))/\sqrt{\varepsilon}$, converges weakly to a specified Gaussian Markov process. Recently, White (in [31] and [32]) has considerably generalized this result, and applied it to a population model.

In the second article [13], Khasminskii treats the special case $G(x) = 0$, and some closely related generalizations. Since $G(x) = 0 \Rightarrow y_\varepsilon(t)$ is constant, it is evident from the above discussion that $x_\varepsilon(t)$ will not have moved appreciably in a time of order $1/\varepsilon$. In [27], Stratonovich argues that the solution, $x_\varepsilon(t)$, will approach a Markov diffusion on intervals with length of order $1/\varepsilon^2$. Assuming, among other conditions, an appropriate form of mixing, Khasminskii defines, in terms of H, a Markov diffusion process, and proves that it is the weak limit of $x_\varepsilon(t)$ on $[0,T/\varepsilon^2]$.

This latter problem has since been treated in considerable detail. See Papanicolaou and Kohler [22], Blankenship and Papanicolaou [3], and, in a more general setting, Cogburn and Hersh [5], as well as some of the references in these articles, for various improvements on Khasminskii's original theorem; in each case establishing the so-called "diffusion limit" on intervals $[0,T/\varepsilon^2]$. The existence of a diffusion limit, of course, precludes the possibility of approximation by a deterministic function, and consequently, such results are of a distinctly different nature from ours.

Averaging on $[0,\infty)$, in the sense of (8), has not yet been explored, and this is our main purpose here (Theorems 1 and 3). Theorem 1 treats a special case - linear systems. Theorem 2 addresses more general systems, establishing a finite interval averaging (similar to Khasminskii [12], Theorem 1.1, and White [31], Theorem 2.1). In Theorem 3, with some additional assumptions, averaging in the general system is extended to $[0,\infty)$.

Here is the setup for Theorem 1. For each ε, $1 \geq \varepsilon > 0$, define

$$\dot{x}_\varepsilon(t,\omega) = A(t,t/\varepsilon,\omega)x_\varepsilon(t,\omega) + d(t,t/\varepsilon,\omega),$$

and

$$\dot{y}_\varepsilon(t) = B_\varepsilon(t)Y_\varepsilon(t) + e_\varepsilon(t)$$

with

$$x_\varepsilon(0,\omega) = y_\varepsilon(0) = x_0 \in R^n,$$

where

(1) For each t, τ, and ω, $A(t,\tau,\omega)$ is an $n \times n$ matrix $\{a_{ij}(t,\tau,\omega)\}$, and $d(t,\tau,\omega)$ is an n-vector $\{d_i(t,\tau,\omega)\}$. For each i and j

(a) a_{ij} and d_i are jointly measurable in their arguments, and

(b) for each ω and $\varepsilon > 0$

$$a_{ij}(t,t/\varepsilon,\omega) \quad \text{and} \quad d_i(t,t/\varepsilon,\omega)$$

are piecewise continuous.

(2) There exist constants c_1 and c_2 such that

$$|A(t,\tau,\omega)| \leq c_1{}^{*}$$

and

$$|d(t,\tau,\omega)| \leq c_2$$

for all t, τ, and ω.

(3) The σ-fields $\mathscr{F}_0^t$ and $\mathscr{F}_t^\infty$ satisfy the Type I mixing condition, where $\mathscr{F}_0^t$ is generated by

$$\{A(s,\tau,\omega), d(s,\tau,\omega): 0 \leq \tau \leq t,\ 0 \leq s < \infty\},$$

and $\mathscr{F}_t^\infty$ is generated by

$$\{A(s,\tau,\omega),\ d(s,\tau,\omega): t \leq \tau < \infty,\ 0 \leq s < \infty\}.$$

(4) $B_\varepsilon(t) = E[A(t,t/\varepsilon)]$ and $e_\varepsilon(t) = E[d(t,t/\varepsilon)]$, and these are piecewise continuous in t for each $1 \geq \varepsilon > 0$. If $\Phi_\varepsilon(t,s)$ is the transition matrix for

$$\dot{y}_\varepsilon(t) = B_\varepsilon(t) y_\varepsilon(t),$$

then there exist positive constants γ and λ such that

$$|\Phi_\varepsilon(t,s)| \leq \gamma e^{-\lambda(t-s)}$$

for all $t \geq s \geq 0$ and $1 \geq \varepsilon > 0$.

---

* $|A|$ refers to the "induced norm", i.e.,

$$|A| = \max_{\substack{x \in R^n \\ |x|=1}} |Ax|.$$

<u>Theorem 1</u>. <u>For all</u> ε <u>sufficiently small</u>, $\sup_{t\geq 0} E|x_\varepsilon(t)|^2 < \infty$, <u>and</u>

$$\lim_{\varepsilon\to 0} \sup_{t\geq 0} E|x_\varepsilon(t) - y_\varepsilon(t)|^2 = 0.$$

<u>Remark</u>: For systems of the form

$$\dot{x}_\varepsilon(t,\omega) = \varepsilon A(t,\omega)x_\varepsilon(t,\omega) + \varepsilon d(t,\omega),$$

the change of variables $t \to t/\varepsilon$ shows that, whenever $E[A(t)]$ is constant, exponential stability of

$$\dot{y}(t) = E[A(t)]y(t)$$

is sufficient for 4.

Now look at some more general systems. Theorems 2 and 3 concern:

$$\begin{aligned} \dot{x}_\varepsilon(t,\omega) &= F(x_\varepsilon(t,\omega),\omega,t,t/\varepsilon) \\ \dot{y}_\varepsilon(t) &= G_\varepsilon(y_\varepsilon(t),t) \\ \dot{x}_\varepsilon(0,\omega) &= y_\varepsilon(0) = x_0 \in R^n \end{aligned} \qquad (17)$$

where

(1) F is jointly measurable with respect to its four arguments, and for all i,j,k, and ω,

$$F_i(x,\omega,t,\tau),\ \frac{\partial}{\partial x_j} F_i(x,\omega,t,\tau),\ \frac{\partial^2}{\partial x_j \partial x_k} F_i(x,\omega,t,\tau)$$

are continuous (in (x,t,τ)).

(2) F is Type II mixing, with $\mathscr{F}_0^t$ the σ-field generated by

$$\{F(x,\omega,s,\tau): 0 \leq \tau \leq t, -\infty < x < \infty, 0 \leq s < \infty\},$$

and $\mathscr{F}_t^\infty$ the σ-field generated by

$$\{F(x,\omega,s,\tau): t \leq \tau < \infty, -\infty < x < \infty, 0 \leq s < \infty\}.$$

(3) $G_\varepsilon(x,t) = E[F(x,\omega,t,t/\varepsilon)]$.

<u>Theorem 2</u>. <u>Assume also that</u>:

(4) <u>There exist continuous functions</u> $B_1(r,t)$, $B_2(r,t)$, <u>and</u> $B_3(r,t)$, <u>such that for all</u> i,j,k, $\tau \geq 0$, <u>and</u> ω:

(a) $|F_i(x,\omega,t,\tau)| \leq B_1(|x|,t)$

(b) $|\frac{\partial}{\partial x_j} F_i(x,\omega,t,\tau)| \leq B_2(|x|,t)$

(c) $\left|\frac{\partial^2}{\partial x_j \partial x_k} F_i(x,\omega,t,\tau)\right| \le B_3(|x|,t)$.

(5) $\sup_{\substack{\varepsilon>0 \\ t\in[0,T]}} |y_\varepsilon(t)| \le K$, _for some constants_ T _and_ K. _Then_

$$\sup_{t\in[0,T]} |x_\varepsilon(t) - y_\varepsilon(t)| \to 0$$

in probability as $\varepsilon \to 0$.

Remarks:

(1) 5 will be satisfied (for every $T > 0$) if, for example,

$$B_1(r,t) \le B(t)(r+1)$$

for some continuous function $B(t)$.

(2) 1 through 5 are not sufficient to guarantee the existence of a solution, $x_\varepsilon(t)$, on $[0,T]$ for every ω. Some realizations of $x_\varepsilon(t)$ may be singular, but this does not affect the conclusion of the theorem (the definition of $x_\varepsilon(t)$ after such a singularity is arbitrary). For an example of this, see [9], section VI.

(3) For linear and certain related systems the theorem remains true using Type III in place of Type II mixing.

With some more conditions, the averaging extends to $[0,\infty)$. Look again at (17) with Assumptions 1, 2, and 3, but restrict B_1, B_2, and B_3 of Assumption 4 to dependence on $|x|$ alone. Further, we require the solution of the deterministic equation, $y_\varepsilon(t)$, to have the following stability property: when perturbed it returns asymptotically to its original trajectory. This is the "physical" meaning of Assumption 6 below. (It does not imply $y_\varepsilon(t) \to y_0$, for some constant y_0.) Since the random solution persistently wanders from the deterministic trajectory, some such stability for the latter is necessary. In bounded linear systems, 6 is equivalent to exponential stability, and in this sense, it is a generalization of Assumption 4 in Theorem 1.

Finally, we assume that $x_\varepsilon(t,\omega)$ is bounded in t,ω, and $\varepsilon > 0$. A more delicate analysis, along the lines of the proof of Theorem 1 (see [9]), would perhaps eliminate this restriction provided that the right-hand side be bounded (at large x) by a linear function of x.

A simple example of an equation to which Theorem 3 applies is

$$\dot{x}_\varepsilon(t,\omega) = -x_\varepsilon(t,\omega) + \frac{\phi(t/\varepsilon,\omega)}{1+x_\varepsilon(t,\omega)^2} \qquad x_\varepsilon(0,\omega) = 0. \tag{18}$$

If ϕ is bounded, Type II mixing, and (say) $E[\phi] = 0$, then

$$\lim_{\varepsilon\to 0} \sup_{t\geq 0} E|x_\varepsilon(t)|^2 = 0.$$

Theorem 3. _In_ (17), _assume_ 1, 2, 3, _and_:

(4) _There exist continuous functions_ $B_1(r)$, $B_2(r)$, _and_ $B_3(r)$, _such that for all_ i,j,k, $t \geq 0$, $\tau \geq 0$, _and_ ω:

(a) $|F_i(x,\omega,t,\tau)| \leq B_1(|x|)$

(b) $|\frac{\partial}{\partial x_j} F_i(x,\omega,t,\tau)| \leq B_2(|x|)$

(c) $|\frac{\partial^2}{\partial x_j \partial x_k} F_i(x,\omega,t,\tau)| \leq B_3(|x|)$.

(5) _There exists a bounded region_ D _such that_

$$x_\varepsilon(t,\omega) \in D$$

for all $\varepsilon > 0$, $t \geq 0$, _and_ ω.

Let $g_\varepsilon(t,s,x)$ _be the solution to_

$$\frac{\partial}{\partial t} g_\varepsilon(t,s,x) = G_\varepsilon(g_\varepsilon(t,s,x),t) \quad g_\varepsilon(s,s,x) = x.$$

(6) _There exists a constant_ K, _and continuous functions_ $M_1(t,s)$ _and_ $M_2(t,s)$, _such that_:

(a) $\sup_{x\in D} |g_\varepsilon(t,s,x)| \leq K$ _for all_ $t \geq s \geq 0$, $\varepsilon > 0$

(b) $K \geq M_1(t,s) \geq \sup_{x\in D} |(\frac{\partial}{\partial x} g_\varepsilon(t,s,x))|$ _for all_ $t \geq s \geq 0$, $\varepsilon > 0$

(c) $M_2(t,s) \geq \sup_{x\in D} |(\frac{\partial^2}{\partial x_i \partial x_j} g_\varepsilon(t,s,x))|$ _for all_ i,j, $t \geq s \geq 0$, $\varepsilon > 0$

(d) $\int_0^t M_1(t,s)ds \leq K$ _for all_ $t > 0$

(e) $\int_0^t M_2(t,s)ds \leq K$ _for all_ $t > 0$.

Then

$$\lim_{\varepsilon\to 0} \sup_{t\geq 0} E|x_\varepsilon(t) - y_\varepsilon(t)|^2 = 0.$$

Remark: 6 can be replaced by 6a, and, for some $\alpha > 0$,

$$\sup_{|y| \leq K} x^T (\frac{\partial}{\partial x} G_\varepsilon(y,t)) x < -\alpha |x|^2 \tag{19}$$

for all $x \in R^n$, $\varepsilon > 0$, and $t \geq 0$ (K as defined in 6a). The remainder of 6 is then a consequence of (19). In the example, (18), (19) holds with any $\alpha < 1$.

5. Stability in Random Homogeneous Linear Systems. The question of averaging relates naturally to the question of stability. If $A(t,\omega)$ is an $n \times n$ matrix of random processes, and

$$B = E[A(t)]$$

is (for now) constant, when does asymptotic stability of the averaged system

$$\dot{y}(t) = By(t) \tag{20}$$

imply some manner of stability for

$$\dot{x}(t,\omega) = A(t,\omega)x(t,\omega)? \tag{21}$$

In this direction, Infante [8] showed that if $A(t,\omega)$ is ergodic, and if for some symmetric positive definite matrix P,

$$E[\text{max eigenvalue}(A^T(t) + PA(t)P^{-1})] < 0, \tag{22}$$

then

$$x(t) \to 0 \quad \text{a.s.}$$

Now suppose (22) is satisfied, and let R be the symmetric positive definite square root of P. Then (22) ⇒

$$E[\max_{\substack{x \in R^n \\ |x|=1}} x^T RA(t)R^{-1} x] < 0 \Rightarrow \max_{\substack{x \in R^n \\ |x|=1}} x^T RBR^{-1} x < 0 \Rightarrow \text{(20) is asymptotically stable.}$$

However, the converse is not true, i.e. the asymptotic stability of (20) does not imply (22). In fact, it is easy to find systems in which (20) and (21) are stable, but (22) is violated. (Recently, Blankenship [2] considerably generalized the condition (22). But still, the above discussion holds, i.e., stability in (20) is not sufficient for the generalized version of (22).)

In a similar vein, Blankenship and Papanicolaou [3] discuss (21) when $A(t)$ is ergodic Markov with compact state space, and the equation is "close" to a related Itô equation, in a suitable sense. In this case, (21) inherits the stability properties of the Itô equation, and the latter are well understood. The point, then, is to utilize the available machinery for Itô equations in analyzing the stability of (21).

Here, we define sufficient conditions on the process $A(t)$ under which the stability of (20) implies that of (21). The main requirements are that: (1) $A(t)$ be bounded, and (2) roughly, $A(t)$ be nearly independent of itself at sufficiently small separations of t. For (2), Type I mixing with "rapidly" decreasing $\rho(\delta)$ will suffice (for example, see Remark 2 below). Notice that $A(t)$ need not be stationary (in fact, B may depend on t).

Specifically, let

$$\dot{x}(t,\omega) = A(t,\omega)x(t,\omega) \qquad x(0,\omega) = x_0 \in R^n.$$

where

(1) $A(t,\omega)$ is an $n \times n$ matrix of real-valued random processes satisfying:

(a) $A(t,\omega)$ is piecewise continuous for each ω,

(b) $|A(t,\omega)| \leq c_1$ for all t and ω, and some constant c_1, and

(c) $B(t) = E[A(t)]$ is piecewise continuous.

(2) The equation

$$\dot{y}(t) = B(t)y(t), \tag{23}$$

is exponentially stable; i.e., if $\Phi(t,s)$ is the transition matrix for (23), then

$$|\Phi(t,s)| \leq \gamma e^{-\lambda(t-s)} \tag{24}$$

for all $t \geq s \geq 0$, and some positive constants γ and λ.

For each $t \geq 0$, define $\mathscr{F}_0^t$ to be the σ-field generated by $\{A(s)\colon 0 \leq s \leq t\}$, and $\mathscr{F}_t^\infty$ to be the σ-field generated by $\{A(s)\colon t \leq s < \infty\}$. For each $\delta > 0$ let

$$\rho(\delta) = \sup_{t \geq 0} \sup_{\substack{A \in \mathscr{F}_0^t \\ B \in \mathscr{F}_{t+\delta}^\infty \\ P(A) > 0}} \frac{1}{P(A)} |v_{t,\delta}(A \times B)|.$$

<u>Theorem 4. There exists an</u> $r_0 > 0$, <u>depending only on</u> $n, c_1, \gamma,$ <u>and</u> λ, <u>such that</u>

$$\min_{\delta \geq 0}(\rho(\delta) + \delta) < r_0 \Rightarrow x(t) \to 0$$

<u>in mean square and almost surely.</u>

Remarks:

(1) Since $|B(t)| \leq c_1$, (24) is equivalent to a variety of seemingly weaker statements, such as

$$\int_{t_0}^{t_1} |\Phi(t,t_0)|dt \leq M$$

for all $t_1 \geq t_0 \geq 0$ and some constant M (see Brockett [4]).

(2) As an example, consider the system

$$\dot{x}_\varepsilon(t,\omega) = A(t/\varepsilon,\omega)x_\varepsilon(t,\omega),$$

where $E[A(t,\omega)]$ is a stability matrix, and A is bounded. If the components of A form a (vector valued) ergodic Markov process, satisfying Doeblin's condition, then A is Type I mixing. The theorem says that

$$x_\varepsilon(t,\omega) \to 0$$

in mean square and almost surely, for all ε sufficiently small.

(3) For situations like the one discussed in Remark 2, the question of whether a.s. convergence holds for every $\varepsilon > 0$ is unresolved. (As far as mean square convergence goes, counter-examples, when ε is too large, are easily constructed.)

Here is a particularly simple example. Let A_1 and A_2 be constant matrices with the property that

$$pA_1 + (1-p)A_2$$

is negative definite, for some $1 > p > 0$. Choose an i.i.d. sequence of A_1's and A_2's with probabilities p and $1 - p$ respectively. For $i = 0,1,\ldots,$ define $A(t,\omega)$ to be the i'th member of the sequence, on the interval $t \in [i,i+1)$. Does $x(t,\omega)$, defined by

$$\dot{x}(t,\omega) = A(t,\omega)x(t,\omega), \tag{25}$$

converge to 0 almost surely? Theorem 4 says that if, instead, we choose $A(t,\omega)$ piecewise constant on the intervals $[i\varepsilon,i\varepsilon+\varepsilon)$, then $x(t,\omega) \to 0$ a.s. for all ε sufficiently small.

Notice that the Kolmogorov zero-one law applies, with the following implication: if $\phi(t,s,\omega)$ is the transition matrix for (25), then either $|\phi(t,0,\omega)| \to 0$ a.s., or

$$p\{|\phi(t,0,\omega)| \to 0\} = 0.$$

6. Continuous Time Stochastic Approximation. Suppose that we observe an R^n valued stochastic process $\phi(t,\omega)$, together with an R^1 valued stochastic process $z(t,\omega)$. We wish to

choose a coefficient vector $x \in R^n$ such that

$$x \cdot \phi(t)$$

approximates, in the sense of minimum mean square error, $z(t)$. If the marginal distribution of $(\phi(t), z(t))$ is independent of t, then the problem has a well-defined solution:

$$\hat{x} = E[\phi\phi^T]^{-1}E[\phi z]$$

(assuming $E[\phi\phi^T]$ is nonsingular*). An obvious adaptation of the discrete time stochastic approximation algorithm for computing $\hat{x}$ (cf. Duda and Hart [8] or Wasan [30]) is

$$\dot{x}_\varepsilon(t,\omega) = \varepsilon\phi(t,\omega)\{z(t,\omega) - \phi(t,\omega)\cdot x_\varepsilon(t,\omega)\}. \tag{26}$$

Averaging gives

$$\dot{y}_\varepsilon(t) = \varepsilon E[\phi z] - \varepsilon E[\phi\phi^T]y_\varepsilon(t),$$

so, clearly, $\lim_{t\to\infty} y_\varepsilon(t) = \hat{x}$ for all $\varepsilon > 0$. If $(\phi(t), z(t))$ is piecewise continuous, Type I mixing, and uniformly bounded in t and ω, then Theorem 1 says

$$\lim_{\varepsilon\to 0} \sup_{t\geq 0} E|x_\varepsilon(t) - y_\varepsilon(t)|^2 = 0,$$

which implies

$$\lim_{\varepsilon\to 0} \overline{\lim_{t\to\infty}} E|x_\varepsilon(t) - \hat{x}|^2 = 0.$$

To get consistency out of (26) (i.e., $x(t) \to \hat{x}$ in some sense), we must let $\varepsilon \to 0$ as $t \to \infty$. In discrete time, the "gain sequence" $1/n$ is commonly used, which suggests that we try

$$\dot{x}(t,\omega) = \frac{1}{t}\,\phi(t,\omega)\{z(t,\omega) - \phi(t,\omega)\cdot x(t,\omega)\} \tag{27}$$

with, say, $x(0,\omega) = x_0 \in R^n$. (See the corollary for more general gain sequences.)

Theorem 5. In (27) assume:

(A1) $\phi(t,\omega)$ and $z(t,\omega)$ are jointly measurable in (t,ω) and, for each ω, piecewise continuous in t.

(A2) $|\phi(t,\omega)| \leq c$ and $|z(t,\omega)| \leq c$ for some constant c, and all t and ω.

(A3) The σ-fields $\mathscr{F}_0^t$ and $\mathscr{F}_t^\infty$ satisfy the Type I mixing condition, where $\mathscr{F}_0^t$ is the σ-field generated by

---

*Equivalently, assuming that

$$c_1\phi_1(t) + c_2\phi_2(t) + \cdots + c_n\phi_n(t) = 0 \quad \text{a.s.} \quad \forall t$$

only when $c_1 = c_2 = \cdots = c_n = 0$.

$$\{\phi(s,\omega), z(s,\omega) : 0 \le s \le t\},$$

and $\mathcal{F}_t^\infty$ *is the* σ*-field generated by*

$$\{\phi(s,\omega), z(s,\omega) : t \le s < \infty\}.$$

(A4) $E[\phi\phi^T]$ *and* $E[\phi z]$ *are independent of* t, *and* $E[\phi\phi^T]$ *is nonsingular*.

Then $x(t) \to \hat{x}$ *in* L^2, *and if*

$$\sum_{n=1}^{\infty} \rho(e^{n\varepsilon}) < \infty \quad \textit{for every} \quad \varepsilon > 0$$

(with $\rho(\delta)$ *as defined in section 2), then the convergence is almost sure as well*.

We can generalize the gain, $1/t$, considerably. For example:

Corollary. Let $\alpha(t)$ be any continuous, strictly decreasing function satisfying

$$\alpha(t) \to 0, \text{ and } \int_0^\infty \alpha(t)dt = \infty.$$

If we replace "$1/t$" by $\alpha(t)$ in (27), then with Assumptions (A1-A4) we still have

$$x(t) \to \hat{x} \quad \text{in} \quad L^2.$$

Furthermore, if for every $\varepsilon > 0$ there exists a summable sequence $\delta_n \downarrow 0$ such that

$$\sum_{n=1}^{\infty} \rho(\delta_n \dot{\beta}(n\varepsilon)) < \infty,$$

then $x(t) \to \hat{x}$ a.s., where $\beta(t)$ is defined by

$$\int_1^{\beta(t)} \alpha(s)ds = t.$$

Proof of the Theorem.

First, make the change of variables $t \to e^t$:

$$\dot{x}(t) = \phi(e^t)z(e^t) - \phi(e^t)\phi(e^t)\cdot x(t) \qquad x(0) = x_0. \tag{28}$$

Henceforth, "$x(t)$" refers to $x(e^t)$, i.e., the solution to (28). We begin by showing

$$\sup_{t \ge 0} E|x(t)|^2 < \infty. \tag{29}$$

(We will establish (29) by a direct analysis of (28). But, if we wished to relax the condition that the averaged system be autonomous (A4), then the approach taken here would not work. One way around this is to use Lemma 1 to establish (29), as is done in the proof of Theorem 1.)

Because of (A2), we can write, for some $k > 0$ and all $0 < \delta < 1$:

$$|x(t) - x(t-\delta)| \leq k\delta|x(t)| + k\delta. \tag{30}$$

Using this:

$$\frac{1}{2}\frac{d}{dt}|x(t)|^2 = z(e^t)\phi(e^t)\cdot x(t) - (\phi(e^t)\cdot x(t))^2$$
$$\leq -(\phi(e^t)\cdot x(t-\delta))^2 + k_1|x(t)| + k_1 + k_1\delta|x(t)|^2$$

for some $k_1 > 0$ and all $0 < \delta < 1$. So, using (A2) to justify the interchange:

$$\frac{1}{2}\frac{d}{dt}E|x(t)|^2 \leq -E(\phi(e^t)\cdot x(t-\delta))^2 \tag{31}$$
$$+ k_1\sqrt{E|x(t)|^2} + k_1 + k_1\delta E|x(t)|^2$$
$$= -\sum_{i,j=1}^{n} E[\phi_i(e^t)\phi_j(e^t)x_i(t-\delta)x_j(t-\delta)]$$
$$+ k_1\sqrt{E|x(t)|^2} + k_1 + k_1\delta E|x(t)|^2.$$

Now $x(t-\delta)$ is $\mathscr{F}_0^{e^{t-\delta}}$ measurable, whereas $\phi(e^t)$ is $\mathscr{F}_{e^t}^{\infty}$ measurable. Consequently, defining $v_{t,\delta}$ as in section 2 and using (12):

$$|E[\phi_i(e^t)\phi_j(e^t)x_i(t-\delta)x_j(t-\delta)]$$
$$- E[\phi_i(e^t)\phi_j(e^t)]E[x_i(t-\delta)x_j(t-\delta)]|$$
$$= \left|\int_{\Omega\times\Omega}\phi_i(e^t,\eta)\phi_j(e^t,\eta)x_i(t-\delta,\omega)x_j(t-\delta,\omega)\,dv_{e^{t-\delta},e^t-e^{t-\delta}}\right|$$
$$\leq 2c^2\rho(e^t-e^{t-\delta})E|x_i(t-\delta)x_j(t-\delta)|$$
$$\leq 2k_2\rho(\tfrac{1}{2}\delta e^t)\{E|x(t)|^2 + E|x(t)| + \delta\}$$

for some $k_2 > 0$ ($0 < \delta < 1 \Rightarrow e^t - e^{t-\delta} \geq \frac{1}{2}\delta e^t$, and $\rho(t)$ is nonincreasing in t). Put this back into (31), letting $b_{ij} = E[\phi_i\phi_j]$:

$$\frac{1}{2}\frac{d}{dt}E|x(t)|^2 \leq -E[\sum_{i,j=1}^{n} b_{ij}x_i(t)x_j(t)]$$
$$+ k_3\sqrt{E|x(t)|^2} + k_3 + k_3\{\delta + \rho(\tfrac{1}{2}\delta e^t)\}E|x(t)|^2$$

for some k_3 sufficiently large. And since, by Assumption (A4),

$$E[\sum_{i,j=1}^{n} b_{ij}x_i(t)x_j(t)] = E\{x^T(t)E[\phi\phi^T]x(t)\} \geq E[\lambda|x(t)|^2]$$

for some $\lambda > 0$, we have

$$\frac{1}{2}\frac{d}{dt} E|x(t)|^2 \leq k_3\{\delta+\rho(\tfrac{1}{2}\ \delta e^t) - \lambda\}E|x(t)|^2 + k_3\sqrt{E|x(t)|^2} + k_3. \qquad (32)$$

Take $\delta < \lambda$ and T so large that

$$\delta + \rho(\tfrac{1}{2}\ \delta e^T) - \lambda < 0.$$

Then $\frac{d}{dt} E|x(t)|^2 < 0$ whenever $t > T$ and $E|x(t)|^2$ is large enough. This proves (29).

Now introduce the averaged equation:

$$\dot{y}(t) = e - By(t) \qquad y(0) = x_0 \qquad (33)$$

where $e = E[\phi z]$ and $B = E[\phi\phi^T]$. If $\psi(t)$ is the transition matrix for $y = -By$, and $g(t,x)$ is the solution to (33) satisfying $g(0,x) = x$, then

$$g(t,x) = \psi(t)x + \int_0^t \psi(t-u)e\,du.$$

Notice that

$$|\psi(t)| \leq \gamma e^{-\lambda t} \qquad (34)$$

for some $\gamma > 0$, where $\lambda > 0$ is the smallest eigenvalue of B.

Apply Lemma 1 with $k(x) = |x-y(t)|^2$: $E|x(t) - y(t)|^2$

$$= 2\int_0^t \int_{\Omega\times\Omega} \{\psi(t-s)x(s,\omega) + \int_s^t \psi(t-u)e\,du - y(t)\}^T\psi(t-s)$$

$$\{\phi(e^s,\eta)z(e^s,\eta) - \phi(e^s,\eta)\phi(e^s,\eta)\cdot x(s,\omega)\}dv_{e^s,0}\,ds.$$

Using (A2), (29), (30), and (34), we can write, for $0 < \delta < 1$,

$$E|x(t) = y(t)|^2 = O(\delta)$$

$$+2\int_\delta^t \int_{\Omega\times\Omega} \{\psi(t-s)x(s-\delta,\omega) + \int_s^t \psi(t-u)e\,du - y(t)\}^T\psi(t-s)$$

$$\{\phi(e^s,\eta)z(e^s,\eta) - \phi(e^s,\eta)\phi(e^s,\eta)\cdot x(s-\delta,\omega)\}dv_{e^s,0}\,d\sigma$$

= (the v integration gives zero for those terms having no ω dependence)

$$2\int_\delta^t \int_{\Omega\times\Omega} \{\psi(t-s)x(s-\delta,\omega)\}^T\psi(t-s)\phi(e^s,\eta)z(e^s,\eta)dv_{e^s,0}\,ds$$

$$-2\int_\delta^t \int_{\Omega\times\Omega} \{\psi(t-s)x(s-\delta,\omega)\}^T\psi(t-s)\phi(e^s,\eta)\phi(e^s,\eta)\cdot x(s-\delta,\omega)dv_{e^s,0}\,ds$$

$$-2\int_\delta^t \int_s^t \int_{\Omega\times\Omega} \{\psi(t-u)e\}^T\psi(t-s)\phi(e^s,\eta)\phi(e^s,\eta)\cdot x(s-\delta,\omega)dv_{e^s,0}\,du\,ds$$

$$+2\int_\delta \int_{\Omega\times\Omega} y(t)^T\psi(t-s)\phi(e^s,\eta)\phi(e^s,\eta)\cdot x(s-\delta,\omega)dv_{e^s,0}\,ds.$$

Since $x(s-\delta)$ is $\mathcal{F}_0^{e^{s-\delta}}$ measurable, and $\phi(e^s)$ and $z(e^s)$ are $\mathcal{F}_{e^s}^{\infty}$ measurable, we can replace $v_{e^s,0}$ in each of the above integrals by $v_{e^{s-\delta},e^s-e^{s-\delta}}$ (these measures agree on $(\Omega \times \Omega, \mathcal{F}_0^{e^{s-\delta}} \times \mathcal{F}_{e^s}^{\infty})$). Next, expand each matrix and vector multiplication, and again apply (12). Together with (A2), (29) and (34), this gives

$$E|x(t) - y(t)|^2 \le k_4\delta + k_4 \int_0^t e^{-\lambda(t-s)}\rho(\tfrac{1}{2}\,\delta e^s)ds \tag{35}$$

(for some $k_4 > 0$)

$$\le k_4\delta + 2k_4 \int_0^T e^{-\lambda(t-s)}ds + k_4 \int_T^t e^{-\lambda(t-s)}\rho(\tfrac{1}{2}\,\delta e^s)ds$$

$$\le k_4\delta + \frac{2}{\lambda}\,k_4 e^{\lambda T}e^{-\lambda t} + \frac{k_4}{\lambda}\,\rho(\tfrac{1}{2}\,\delta e^T).$$

To prove the first assertion of the theorem, take $T = t/2$ and $\delta = 1/t$ to get

$$\lim_{t\to\infty} E|x(t) - y(t)|^2 = 0,$$

and observe that $y(t) \to E[\phi\phi^T]^{-1}E[\phi z]$.

For almost sure convergence it is sufficient (because of (30)) to show

$$|x(n\varepsilon) - y(n\varepsilon)| \to 0 \quad \text{a.s.} \tag{36}$$

for every $\varepsilon > 0$. In (35) take $t = n\varepsilon$, $T = \frac{1}{2}\,n\varepsilon$, and $\delta = \frac{1}{n^2}$ to get

$$E \sum_{n=1}^{\infty} |x(n\varepsilon) - y(n\varepsilon)|^2 < \infty,$$

which implies (36). Q.E.D.

Proof of the Corollary. Start with the change of variables $t \to \beta(t)$ (instead of $t \to e^t$), and proceed exactly as above. This leads to

$$\frac{1}{2}\frac{d}{dt}E|x(t)|^2 \le k_3\{\delta + \rho(\beta(t) - \beta(t-\delta)) - \lambda\}E|x(t)|^2 \tag{37}$$

$$+ k_3\sqrt{E|x(t)|^2} + k_3$$

in place of (32). From the definition of $\beta(t)$:

$$\delta = \int_{\beta(t-\delta)}^{\beta(t)} \alpha(s)ds \le \alpha(\beta(t-\delta))(\beta(t) - \beta(t-\delta))$$

$$\Rightarrow \beta(t) - \beta(t-\delta) \ge \frac{\delta}{\alpha(\beta(t-\delta))} = \delta\dot{\beta}(t-\delta).$$

Notice that $\dot\beta(t) \uparrow \infty$ as $t \to \infty$, since $\dot\beta(t) = 1/\alpha(\beta(t))$ and $\beta(t) \uparrow \infty$. Hence

$$\rho(\beta(t) - \beta(t-\delta)) \le \rho(\delta\dot\beta(t-1)) \tag{38}$$

(for $0 < \delta < 1$). Now put (38) into (37) and repeat the argument used with (32), to establish (29).

Again, proceed as in the proof of the theorem, to get

$$E|x(t) - y(t)|^2 \le k_4\delta + \frac{2}{\lambda} k_4 e^{\lambda T} e^{-\lambda t} + \frac{k_4}{\lambda} \rho(\delta\dot\beta(T-1)) \tag{39}$$

in place of (35). For L^2 convergence, put $T = t/2$ and take $\delta(t) \downarrow 0$ such that $\delta(t)\dot\beta(t/2-1) \to \infty$. For almost sure convergence, fix $\varepsilon > 0$ and choose a summable sequence $\delta_n \downarrow 0$ such that

$$\sum_{n=1}^{\infty} \rho(\delta_n \dot\beta(\frac{n\varepsilon}{2} - 1)) < \infty.$$

From (39), with $T = \frac{n\varepsilon}{2}$ and $t = n\varepsilon$,

$$E|x(n\varepsilon) - y(n\varepsilon)|^2 \le k_4\delta_n + \frac{2}{\lambda} k_4 e^{-\frac{\lambda n\varepsilon}{2}} + \frac{k_4}{\lambda} \rho(\delta_n \dot\beta(\frac{n\varepsilon}{2} - 1)).$$

Hence, $E \sum_{n=1}^{\infty} |x(n\varepsilon) - y(n\varepsilon)|^2 < \infty$, and, as with the proof of the theorem, this is enough for $x(t) \to \hat{x}$ a.s. Q.E.D.

Many of the stochastic approximation algorithms can be viewed as modified versions of the gradient descent procedure for finding an extremum of a multivariate function. In the regression problem, for example, the "criterion" for fixed x, was $(z(t) - \phi(t)\cdot x)^2$, and we sought to minimize the function

$$H(x) = E[(z-\phi\cdot x)^2].$$

The gradient descent solution is to solve

$$\dot{x} = -\nabla H(x)$$

which, of course, leads to $x(t) \to \hat{x}$. But we do not have

$$-\nabla H(x) = E[2\phi(z-\phi\cdot x)]$$

available - it cannot be observed. On the other hand, we can observe $\phi(z-\phi\cdot x)$, but

$$\dot{x} = \phi(z-\phi\cdot x)$$

does not give convergence. With a slight modification we arrive at (27), which works.

This leads us to the study of

$$\dot{x}(t,\omega) = \frac{1}{t} F(x(t,\omega),\omega,t) \qquad x(1) \in R^n \tag{40}$$

where, for some function $H(x)$ (which we think of as the mean of a criterion function),

$$E[F(x,\omega,t)] = -\nabla H(x). \tag{41}$$

(From a different point of view, our interest may be in finding a critical point of some deterministic function $H(x)$. We make observations of $-\nabla H(x)$, but these are disturbed in such a way that we observe, instead, a random variable $F(x,\omega,t)$, related to $-\nabla H(x)$ by (41).)

Given (40) and (41), we will assume:

(A1) (i) F is jointly measurable in its three arguments,

(ii) for each ω, F is of class C^2 in x at every t, and

(iii) there exists a continuous function, $B(x)$, which bounds $|F_i(x,\omega,t)|$, $|\frac{\partial}{\partial x_j} F_i(x,\omega,t)|$, and $|\frac{\partial^2}{\partial x_j \partial x_k} F_i(x,\omega,t)|$ for all i,j,k, $t \geq 0$, x, and ω.

(A2) For each ω, (40) has a unique absolutely continuous solution, $x(t,\omega)$, on $[0,\infty)$ satisfying

$$\sup_{t\geq 0} |x(t,\omega)| < \infty. \tag{42}$$

(A3) The set of critical values of H contains no open intervals.*

(A4) Let $G(x) = E[F(x,\omega,t)]$. The equation $\dot{y} = G(y)$ $y(0) = y_0$ has a solution on $[0,\infty)$ for every y_0.

(A5) The σ-fields $\mathscr{F}_0^t$ and $\mathscr{F}_t^\infty$ satisfy the Type II mixing condition, where $\mathscr{F}_0^t$ is generated by

$$\{F(x,\omega,s): 0 \leq s \leq t, -\infty < x < \infty\},$$

and $\mathscr{F}_t^\infty$ is generated by

$$\{F(x,\omega,s): t \leq s < \infty, -\infty < x < \infty\}.$$

Let $D = \{x: H(x) = 0\}$, and let $d(S_1,S_2)$ denote the distance between sets S_1 and S_2 (i.e., $\inf\{|x_1-x_2|: x_1 \in S_1, x_2 \in S_2\}$). Using the σ-fields defined in (A5), define $\rho(\delta)$ and $v_{t,\delta}$ as in section 2.

---

*y is a critical value of H if, for some x, $H(x) = y$ and $\nabla H(x) = 0$. Notice that (A1) implies that H is of class C^3. The reader can convince himself that, given (A1), counter-examples to (A3) are extremely hard to come by. If H is of class C^n, then the condition follows from the Morse-Sard theorem.

<u>Theorem 6</u>. <u>If for every</u> $\varepsilon > 0$

$$\sum_{n=1}^{\infty} \rho(e^{\varepsilon n}) < \infty, \tag{43}$$

<u>then</u> $d(x(t),D) \to 0$ <u>almost surely</u>.

<u>Remarks</u>:

(1) The assumptions can be trimmed considerably, but then the proof becomes more technical. Our main purpose is to illustrate the averaging approach to stochastic approximation. Usually, specific examples are better handled individually, as in Theorem 5.

(2) (42) may be obvious from the system at hand, or, it can be imposed on the equation - by a modification of F which turns x inward at a preassigned boundary. Nevertheless, as it stands (42) is unsatisfying. The proposition below (immediately following the proofs of the theorem and corollary) gives sufficient conditions for (42), but here again the conditions are far from the most general.

Just as with Theorem 5, more general gains present no new problems:

<u>Corollary</u>. Take $\alpha(t)$ and $\beta(t)$ as in the corollary to Theorem 5, and assume that for every $\varepsilon > 0$ there exists a summable sequence $\delta_n \downarrow 0$ such that

$$\sum_{n=1}^{\infty} \rho(\delta_n \dot{\beta}(n\varepsilon)) < \infty.$$

Then, with "1/t" replaced by $\alpha(t)$ in (40), we still have $d(x(t),D) \to 0$ a.s.

<u>Proof of Theorem 6</u>. In (40) make the change of variables $t \to e^t$:

$$\dot{x}(t,\omega) = F(x(t,\omega),\omega,e^t) \qquad x(0) = x_0, \tag{44}$$

and let $x(t)$, from now on, refer to the solution to (44).

We will show that for every $\xi > 0$

$$\lim_{n\to\infty} d(x(n\xi),D) = 0 \qquad \text{a.s.} \tag{45}$$

This is enough, because for almost every ω, $\dot{x}(t,\omega)$ is uniformly bounded in t (follows from (A1) and (A2)).

Now fix $\varepsilon > 0$, Since we are dealing with a discrete sequence of random variables, it is convenient to define

$$x_n = x(n\xi) \qquad n = 0,1,... \tag{46}$$

We know that the solution to the averaged equation, $\dot{y} = G(y) = -\nabla H(y)$, moves in the right direction - the direction in which H decreases most rapidly. So, we will want to relate the changes in x_n to the flow of the averaged equation. Define $g(t,x)$ by

$$\frac{\partial}{\partial t} g(t,x) = G(g(t,x)) \qquad g(0,x) = x$$

and, for $n = 1,2,...,$ define

$$y_n = g(\xi, x_{n-1}). \tag{47}$$

The proof of (45) proceeds in two steps. The first is a purely deterministic argument: if for some bounded sequence $\{x_n\}$, $|x_n - y_n| \to 0$, then

$$d(x_n, D) \to 0. \tag{48}$$

The second step is to use Lemma 1 to prove that $|x_n - y_n| \to 0$ a.s. (The latter being suggested by the fact that the mixing rate of F, in (44), increases with increasing t.)

Let $\{x_n\}$ be a sequence of vectors in R^n such that

$$|x_n| \leq r$$

for some $r > 0$, and

$$|x_n - y_n| \to 0,$$

where y_n is defined by (47). Define

$$R = \{x: |x| \leq r\},$$

and, for any ℓ,

$$L_\ell = \{x: H(x) \leq \ell\}.$$

If ℓ is in the range of H over the domain R, and if ℓ is not a critical value of H, then, we claim,

$$x_n \in L_\ell \quad \text{infinitely often} \Rightarrow$$
$$x_n \in L_\ell \quad \forall n \ \text{ sufficiently large.}$$

In other words, x_n can leave L_ℓ at most a finite number of times. The reason is that x_n must eventually follow y_n very closely, and

$$H(y_n) \leq H(x_{n-1}).$$

More precisely, let

$$\varepsilon = d(H^{-1}(\ell), g(\xi, L_\ell \cap R)).$$

Then $\varepsilon > 0$ since

$$g(\xi, L_\ell \cap R) \subset \text{interior of } L_\ell$$

and $g(\xi, L_\ell \cap R)$ is compact ($g^{-1}(\xi,\cdot)$ is continuous). Choose N so large that $n > N \Rightarrow |x_n - y_n| < \varepsilon$. Then $n > N$ and $x_n \in L_\ell \Rightarrow$

$$d(x_{n+1}, g(\xi, L_\ell \cap R)) < \varepsilon$$

(since $y_{n+1} = g(\xi, x_n) \in g(\xi, L_\ell \cap R)) \Rightarrow x_{n+1} \in L_\ell$.

Now to prove (48), suppose that $\{x_n\}$ has a limit point x^* in the complement of D. Choose a neighborhood N of x^* such that

$$\sup_{x \in N} H(g(\xi,x)) < \inf_{x \in N} H(x),$$

and then choose an ℓ such that ℓ is not a critical value and

$$\ell \in [\sup_{x \in N} H(g(\xi,x)), \inf_{x \in N} H(x)].$$

Eventually, x_n must enter L_ℓ after each visit to N. But we can have already established that x_n cannot leave L_ℓ infinitely often, which contradicts the existence of x^*.

Now we turn to the probabilistic aspect of the proof. For the random sequence $\{x_n\}$, defined in (46), we wish to show

$$|x_n - y_n| \to 0 \quad \text{a.s.} \tag{49}$$

Fix an $r > 0$ and look at the set of all ω such that

$$\sup_{t \ge 0} |x(t)| \le r. \tag{50}$$

For these paths, with the possible exception of a set of probability zero, we will prove that $|x_n - y_n| \to 0$. Because of the Assumption (A2), and because r is arbitrary, this is enough for (49).

Let

$$s = \sup_{\substack{|x| \le r \\ t \in [0,\xi]}} |g(t,x)|.$$

Without changing $F(x,\omega,t)$ when $|x| \le s$, modify F so that F is zero whenever $|x| \ge s + 1$ but F still satisfies the Assumptions (A1), (A2), A4), and (A5) (e.g., multiply F by a smooth function of x which is 1 when $|x| \le s$ and 0 when $|x| \ge s + 1$). Evidently, if $x(t,\omega)$ satisfies (50) for some ω, then $x(t,\omega)$ and the corresponding sequence $\{y_n(\omega)\}$ are unaffected by this modification. The point is that we can assume

$$\begin{aligned} &\sup_{t \ge 0} |x(t)| \le s + 1 \quad \text{a.s., and} \\ &\sup_{n} |y_n| \le s + 1 \quad \text{a.s.} \end{aligned} \tag{51}$$

and it will suffice to show $|x_n - y_n| \to 0$ a.s. under these assumptions.

With initial conditions given at $t = (n-1)\xi$, and with $K(x,\omega) = |x - y_n(\omega)|^2$, apply Lemma 1 to (44) and proceed as in the proof of Theorem 5:

$$E|x_n - y_n|^2 = \int_{(n-1)\xi}^{n\xi} \int_{\Omega\times\Omega} (\frac{\partial}{\partial x} K(g(n\xi - s, x(s,\omega)),\omega)) \cdot F(x(s,\omega),\eta,e^s)\, dv_{e^s,0}\, ds.$$

For any $n \geq 2$ and $0 < \delta < 1$:

$$E|x_n - y_n|^2$$

$$= \int_{(n-1)\xi}^{n\xi} \int_{\Omega\times\Omega} (\frac{\partial}{\partial x} K(g(n\xi - s, x(s-\delta,\omega)),\omega)) \cdot \tag{I}$$

$$F(x(s-\delta,\omega),\eta,e^s)\, dv_{e^s,0}\, ds$$

$$+ \int_{(n-1)\xi}^{n\xi} \int_{\Omega\times\Omega} \{ (\frac{\partial}{\partial x} K(g(n\xi - s, x(s,\omega)),\omega)) \cdot F(x(s,\omega),\eta,e^s) \tag{II}$$

$$- (\frac{\partial}{\partial x} K(g(n\xi - s, x(s-\delta,\omega)),\omega)) \cdot F(x(s-\delta,\omega),\eta,e^s) \} dv_{e^s,0}\, ds.$$

In (I), replace $v_{e^s,0}$ by $v_{e^{s-\delta},e^s-e^{s-\delta}}$ and apply (51), (A1), and (11) to get (for some constant c):

$$\text{"I"} \leq c \sup_{s\in[(n-1)\xi,n\xi]} \rho(e^s - e^{s-\delta}) = c\rho(e^{n\xi} - e^{n\xi-\delta}) \leq c\rho(e^{n\xi} \frac{\delta}{2}).$$

For (II) we have

$$\text{"II"} \leq \int_{(n-1)\xi}^{n\xi} \int_{\Omega\times\Omega} |x(s,\omega) - x(s-\delta,\omega)|$$

$$\sup_{|x|\leq s+1} |\nabla\{ (\frac{\partial}{\partial x} K(g(n\xi - s, x),\omega)) \cdot F(x,\eta,e^s)\}|\, d|v|_{e^s,0}\, ds.$$

(A1) and (51) imply that, uniformly in ω,

$$|x(s,\omega) - x(s-\delta,\omega)| = O(\delta)$$

and

$$\sup_{|x|\leq s+1} |\nabla\{ (\frac{\partial}{\partial x} K(g(n\xi - s, x),\omega)) \cdot F(x,\eta,e^s)\}| = O(1).$$

Hence,

$$E|x_n - y_n|^2 = O(\rho(e^{n\xi} \frac{\delta}{2})) + O(\delta).$$

Take $\delta = 1/n^2$ and apply (43) to get

$$\sum_{n=2}^{\infty} E|x_n - y_n|^2 < \infty, \tag{52}$$

which completes the proof. Q.E.D.

Proof of the Corollary. The same, except that we begin with the change of variables $t \to \beta(t)$, and then use $v_{\beta(t),0}$ in place of $v_{e^t,0}$, $\rho(\beta(t) - \beta(t-\delta))$ in place of $\rho(e^t - e^{t-\delta})$, and $\rho(\delta\dot{\beta}(t-1))$ in place of $\rho(\frac{\delta}{2} e^t)$ (see proof of the previous corollary), to get

$$E|x_n - y_n|^2 = O(\rho(\delta\dot{\beta}(n\xi-1))) + O(\delta).$$

Now, for (52), take δ_n summable such that $\rho(\delta_n\dot{\beta}(n\xi-1))$ is summable. Q.E.D.

One way to establish (42) is to use Lemma 1, as is illustrated in:

Proposition. With reference to (40) and (41) assume:

(A1) (i) F is jointly measurable in its three arguments,

(ii) for each ω, F is of class C^2 in x at every t, and

(iii) there exists a constant, B, which bounds

$$|F_i(x,\omega,t)|,\ |\frac{\partial}{\partial x_j} F_i(x,\omega,t)|,\ \text{and}$$

$$|\frac{\partial^2}{\partial x_j \partial x_k} F_i(x,\omega,t)| \quad \text{for all } i,j,k,t\geq 0,x, \text{ and } \omega.$$

(A2) For each ω, (40) has a unique absolutely continuous solution, $x(t,\omega)$, on $[0,\infty)$.

(A3) $|H(x)| \to \infty$ as $|x| \to \infty$.

(A4) The σ-fields $\mathscr{F}_0^t$ and $\mathscr{F}_t^\infty$ satisfy the Type II mixing condition, where $\mathscr{F}_0^t$ is generated by

$$\{F(x,\omega,s):\ 0 \leq s \leq t,\ -\infty < x < \infty\},$$

and $\mathscr{F}_t^\infty$ is generated by

$$\{F(x,\omega,s):\ t \leq s < \infty,\ -\infty < x < \infty\}.$$

If $\rho(\delta) = O((\log \delta)^{-4})$ as $\delta \to \infty$ * then

$$\sup_{t\geq 0}|x(t)| < \infty \quad \text{a.s.}$$

Proof. Begin as in the proof of Theorem 6: make the change of variables $t \to e^t$ to get (44). Obviously, it is enough to show

$$\sup_{n=0,1,\ldots} |x_n| < \infty \qquad \text{a.s.} \tag{53}$$

*Many Markov processes, for example, are Type I mixing with exponentially decreasing mixing rates (see section 2).

where $x_n = x(n)$. With $G(x) = E[F(x,\omega,t)] = -\nabla H(x)$, the averaged equation is $\dot{y}(t) = G(y(t))$. Define $g(t,x)$ by $\frac{\partial}{\partial t} g(t,x) = G(g(t,x))$ $g(0,x) = x$ and, for $n = 1,2,\ldots,$ define $y_n = g(1,x_{n-1})$.

In the usual way, apply Lemma 1. Everything is bounded, and just as in Theorem 6 we get

$$E|x_n-y_n|^2 = O(\rho(e^n \tfrac{\delta}{2})) + O(\delta).$$

If $\delta = \frac{1}{n^4}$, then

$$E|x_n-y_n|^2 = O(\rho(e^n \frac{1}{2n^4})) + O(\frac{1}{n^4}) = O(\frac{1}{n^4}).$$

From Chebyshev's inequality,

$$P(|x_n-y_n| > \frac{1}{n^{4/3}}) = O(\frac{1}{n^{4/3}}),$$

and then, from the Borel-Cantelli lemma,

$$\sum_{n=1}^{\infty} |x_n-y_n| < \infty \quad \text{a.s.}$$

Finally, using (A1), we have

$$H(x_n) \leq H(y_n) + B|x_n-y_n| \leq H(x_{n-1}) + B|x_n-y_n|$$

$$\Rightarrow H(x_n) \leq H(x_0) + B \sum_{k=1}^{n} |x_k-y_k|$$

$$\Rightarrow \sup_{n\geq 1} H(x_n) < \infty \quad \text{a.s.} \Rightarrow (53). \qquad \text{Q.E.D.}$$

The approach applies, as well, to other problems of stochastic approximation. For example, a continuous time version of the Kiefer-Wolfowitz procedure (see [14] for the original formulation) can be treated in very much the same way. The main difference is that the averaged equation is nonautonomous ($G = G(x,t)$), but this presents no new difficulties (for example, see sections 4 and 5). The basic recipe is the same: use Lemma 1 to show that the solutions to the random equation converge, in some sense, to those of the averaged equation, and check that the (determinsitic) averaged equation approaches the desired limit.

Wasan [30] is a good general reference (and source for references) on stochastic approximation, but the literature on continuous time procedures, in particular, is sparse. An earlier paper somewhat related to the present study, is Driml and Nedoma [7]. However, the assumptions there are extremely specialized. More closely related is an article by Sakrison [25], which treats a continuous Kiefer-Wolfowitz procedure, modified to insure boundedness, under a mixing-like assumption. There has also been some work done on continuous stochastic approximation when the disturbances are Gaussian white noise, leading to an analysis of convergence in certain Itô-type stochastic differential equations (for a recent article, and further references, see Sorour [26]).

However, what we wish to emphasize is not the continuous time aspect of the approach, but its ability to handle dependent (mixing) and correlated observations. In this regard, the results here are closely related to some recent work by Kushner ([15] and [16] and references therein) and by Ljung (cf. [17], [18], and [19]). It is difficult to compare the various approaches (and not necessarily appropriate - since each should be taken as a possible technique for handling specific cases). For example, the assumptions of Theorem 6 (if reformulated in discrete time) are stronger than those of the Theorem in [15],* but the conclusion is also stronger: L^2 and almost sure convergence as opposed to convergence in probability. On the other hand, Ljung, in [17] and [19], obtains almost sure convergence in a setting similar to that of Theorem 6, but in this case the assumptions are too different to be meaningfully related. What is common to each of these approaches is the demonstration that stochastic approximation procedures are effective in a far broader context than that in which they were originally formulated.

REFERENCES

[1] Billingsley, P. (1968) Convergence of Probability Measures, John Wiley and Sons, New York.

[2] Blankenship, G. (1977) Stability of linear differential equations with random coefficients, IEEE Trans. Automatic Control, AC-22, 834-838.

[3] Blankenship, G. and Papanicolaou, B.C. (1978) Stability and control of stochastic systems with wide-band noise disturbances I, SIAM J. Appl. Math. 34, 437-476.

[4] Brockett, R.W. (1970) Finite Dimensional Linear Systems, John Wiley and Sons, New York.

*Although, here we allow the noise to enter the system in a more general manner.

[5] Cogburn, R. and Hersh, R. (1973) Two limit theorems for random differential equations, Indiana U. Math. J., 22, 1067-1089.

[6] Doob, J.L. (1953) Stochastic Processes, John Wiley and Sons, New York.

[7] Driml, M. and Nedoma, J. (1959) Stochastic approximations for continuous random processes, Transactions of the Second Prague Conference on Information Theory, Statistical Decision Functions and Random Processes, Prague, 145-158.

[8] Duda, R.O. and Hart, P.E. (1973) Pattern Classification and Scene Analysis, John Wiley and Sons, New York.

[9] Geman, S. Some averaging and stability results for random differential equations, SIAM J. Appl. Math. (to appear).

[10] Hartman, P. (1964) Ordinary Differential Equations, John Wiley and Sons, New York.

[11] Infante, E.F. (1968) On the stability of some linear non-autonomous random systems, J. of Appl. Mechanics, 35, 7-12.

[12] Khasminskii, R.Z. (1966) On stochastic processes defined by differential equations with a small parameter, Theor. Prob. Appl., 11, 211-228.

[13] Khasminskii, R.Z. (1966) A limit theorem for the solutions of differential equations with random right-hand sides. Theor. Prob. Appl., 11, 390-406.

[14] Kiefer, J. and Wolfowitz, J. (1952) Stochastic estimation of the maximum of a regression function, Ann. Math. Stat., 23, 462-466.

[15] Kushner, H.J. (1977) Convergence of recursive adaptive and identification procedures via weak convergence theory, IEEE Trans. Automatic Control, AC-22, 921-930.

[16] Kushner, H.J. (1977) General convergence results for stochastic approximations via weak convergence theory, J. of Math. Analysis and Appl., 61, 490-503.

[17] Ljung, L. (1974) Convergence of recursive stochastic algorithms, Report 7403, Dept. of Automatic Control, Lund Institute of Technology, Sweden.

[18] Ljung, L. (1977) Analysis of recursive stochastic algorithms, IEEE Trans. Automatic Control, AC-22, 551-575.

[19] Ljung, L. (1978) Stong convergence of a stochastic approximation algorithm, Ann. of Stat., 6, 680-696.

[20] Lybrand, E.R. (1975) Method of averaging for stochastic systems, Ph.D. Thesis, Electrical Engineering Department, Polytechnic Institute of Brooklyn.

[21] Mitropolsky, Iu.A. (1967) Averaging method in non-linear mechanics, Int. J. of Non-Linear Mechanics, 2, 69-96.

[22] Papanicolaou, G.C. and Kohler, W. (1974) Asymptotic theory of mixing stochastic ordinary differential equations, Comm. Pure Appl. Math., 27, 641-668.

[23] Rosenblatt, M. (1956) A central limit theorem and a strong mixing condition, Proc. Nat. Acad. Sci., 42, 43-47.

[24] Rozanov, Yu.A. (1967) Stationary Random Processes, Holden-Day, San Francisco.

[25] Sakrison, D.J. (1964) A continuous Kiefer-Wolfowitz procedure for random processes, Ann. Math. Stat., 35, 590-599.

[26] Sorour, E.S. (1977) Continuous stochastic approximation procedure for evaluating the point at which the regression function stops to be non-positive, Kybernetika, 13, 450-464.

[27] Stratonovich, R.L. (1963,I), (1967,II) Topics in the Theory of Random Noise, Gordon and Breach, New York.

[28] Volkonskii, V.A. and Rozanov, Yu.A. (1959) Some limit theorems for random functions I, Theor. Probability Appl., 4, 178-197.

[29] Vrkoc, I. (1966) Extension of the averaging method to stochastic equations, Czech. Math. J., 16, 518-540.

[30] Wasan, M.T. (1969) Stochastic Approximation, Cambridge University Press, Cambridge.

[31] White, B.S. (1976) Some limit theorems for stochastic delay-differential equations, Comm. Pure Appl. Math., 29, 113-141.

[32] White, B.S. (1977) The effects of a rapidly-fluctuating random environment on systems of interacting species, SIAM J. Appl. Math., 32, 666-693.

Stochastic Approximation and Nonlinear Operator Equations*

R. Kannan

Department of Mathematics
University of Texas at Arlington, Arlington, Texas 76019

1. Introduction
2. Convex Functionals and Existence of Minima
3. Maximal Monotone Operators
4. Algorithms for Approximate Solutions
5. The Stochastic Case
6. Convergence of the Algorithm: Discrete Time
7. Continuous Time Stochastic Approximation
8. Tikhonov Regularization: Stochastic Case
References

*Research partially supported by University of Texas at Arlington Research Grant.

1. Introduction. We consider the nonlinear optimization problem of minimizing a nonlinear functional $f(x)$ over a real Hilbert space H. For the sake of simplicity, we restrict ourselves to the case of the unconstrained problem. Thus we are trying to locate $x^* \in H$ at which $f(x)$ attains its minimum. The main difficulty is that we cannot calculate $f(x)$. All we can obtain are observations of the type $f(x) + \varepsilon(x)$ where $\varepsilon(x)$ is the random disturbance. We would like to generate a sequence $\{x_n\}$ by methods analogous to the deterministic case (where we can calculate $f(x)$) such that $\{x_n\}$ converges in a probabilistic sense to x^*. As is true in a wide variety of practical situations, x^* is not necessarily unique and thus we would like to obtain sufficient conditions which guarantee that x_n converges with probability one to an element of the set of points where f attains its minimum.

In the deterministic case, if $f(x)$ is either differentiable or the subdifferential $Df(x)$ (to be defined later) exists, the problem of minimizing $f(x)$ reduces to the nonlinear operator equation $Df(x) = 0$ where Df is a nonlinear operator defined over H. It is known that under suitable continuity hypotheses Df is a maximal monotone operator and thus the problem of finding a sequence $\{x_n\}$ converging to a minimum point of $f(x)$ can be looked at as a problem of finding approximate solutions of the operator equation $Df(x) = 0$. We note here that if the functional f was itself random then the corresponding operator equation would be a random operator equation $Df(x,\omega) = 0$.

Thus it is natural to expect that the known algorithms for approximating the solutions of $Df(x) = 0$ should be adaptable to the stochastic case. It is our purpose in this paper to discuss this point in detail. Two of the methods that are used for finding approximate solutions in the deterministic case are the regularization method of Tikhonov and the algorithm analogous to the gradient direction method. We will discuss the ramifications in the stochastic case later. However we make the following remark. The gradient direction method gives rise to the following algorithm in the deterministic case: consider the sequence of elements x_n generated by

$$x_{n+1} = x_n - a_n Df(x_n).$$

Thus the analogous algorithm in the stochastic case would be to consider the sequence of random variables

$$x_{n+1}(\omega) = x_n(\omega) - a_n Df(x_n,\omega).$$

But this is similar to the stochastic approximation procedure of Robbins-Monro and in fact many of the variants of the Robbins-Monro procedure may be viewed as modifications of the gradient direction method.

An alternative to "discrete time" in the gradient direction method would be to consider the continuous analogue, i.e., the case of continuous time stochastic approximation. In fact, in the deterministic case, continuous algorithms for minimizing $f(x)$ can be written in a broad sense in the form

$$\dot{x}(t) = -T(x,t), \quad x(0) = x_0$$

where $T(x,t)$ is a continuous function with values in H satisfying the condition

$$\langle T(x,t), Df(x,t) \rangle > 0;$$

i.e., $T(x,t)$ forms an acute angle with the gradient direction: thus one is led to the corresponding continuous time stochastic approximation procedure

$$\dot{x}(t,\omega) = -tD_x f(x,t,\omega), \quad x(0) = x .$$

The question now is to show $x(t,\omega) \to x^*$ a.s. as $t \to \infty$.

In the present paper we do not go into the technical details of and proofs. We restrict ourselves to outlining the pertinent concepts and results from the deterministic and stochastic theories and the directions in which the problem of approximate solutions of the random operator equations that arise may be approached. In this sense this paper may be viewed as an attempt to point out the adaptability of deterministic optimization techniques to the stochastic case.

In Sections 2, 3 and 4 of this paper we outline some of the concepts and results from the deterministic case, while in Section 5 the questions that arise in the stochastic case are introduced and a brief introduction to the theory of random operator equations is presented. Section 6 deals with the stochastic analogue of the gradient direction method and the connection with the Robbins-Monro procedure. Convergence results for $x_n(\omega)$ are presented. The continuous analogue is then discussed in Section 7. We conclude the paper with some discussions of the stochastic analogue of the Tikhonov regularization method.

2. Convex Functionals and Existence of Minima. Let H be a real Hilbert space and $f(x)$ a real functional defined on H. f is

said to be convex if for every $x, y \in H$ and $\alpha \in (0,1)$ we have

$$f(\alpha x + (1-\alpha)y) \leq \alpha f(x) + (1-\alpha)f(y).$$

If in addition for $x \neq y$ the strict inequality holds, we say that f is strictly convex. A sequence $\{x_n\}$, $x_n \in H$ is said to be a minimizing sequence for f if $f(x_n) \to \inf_{x \in H} f(x)$.

It can be seen that if $H = R^n$, the condition

$$\lim_{t \to +\infty} f(tx) = +\infty \tag{1}$$

for any $x \in H$, $x \neq 0$ guarantees the existence of a minimum of f in H and every minimizing sequence converges to the minimum of f if f is strictly convex. Moreover the convexity of f implies its continuity. However if H is infinite dimensional a convex functional which satisfies (1) need not be even bounded from below. Even if f is bounded from below, f does not necessarily attain its minimum if H is infinite dimensional. Simple examples may be constructed in ℓ_2. Finally, continuity of f is not necessarily true in the case of infinite dimensional H.

We say that f is weakly lower semicontinuous at a point $x_0 \in H$ if for every sequence $\{x_n\}$ converging weakly to x_0 we have

$$f(x_0) \leq \underline{\lim}_{n \to \infty} f(x_n).$$

Proposition 2.1. If the functional f is weakly lower semicontinuous on a bounded weakly closed set S in a reflexive Banach space, then f is bounded from below.

Proposition 2.2. A weakly lower semicontinuous functional f which satisfies

$$\overline{\lim}_{\|x\| \to \infty} f(x) = +\infty$$

is bounded from below and attains its minimum.

Definition. The operator $F: H \to H$ defined by the formula

$$\lim_{t \to 0} \frac{1}{t}\left[f(x+th) - f(x)\right] = \langle F(x), h \rangle$$

for any $h \in H$, is called the gradient of f and $Df(x,h) = \langle F(x), h \rangle$ is called the Gateaux differential of f.

The following results show the connection between the weak lower semicontinuity of f and $Df(x,h)$.

Proposition 2.3. If at a point x_0 we have

$$f(x) - f(x_0) - Df(x_0, x-x_0) \geq 0$$

for every x belonging to a neighborhood of x_0, f is weakly lower semicontinuous at x_0.

Proposition 2.4. If for each $x \in H$ (fixed) the differential $Df(x,h)$ is continuous in h, then $f(x)$ is weakly lower semicontinuous.

Proposition 2.5. If the gradient $F(x)$ admits a Gateaux derivative $F'(x)$ and if

$$\langle F'(x)h, h \rangle \geq 0$$

then $f(x)$ is weakly lower semicontinuous.

For a convex functional with a Gateaux differential we have:

Proposition 2.6. A necessary and sufficient condition that a convex functional $f(x)$ have a minimum at a point $x_0 \in H$ is

$$\text{grad } f(x_0) = F(x_0) = 0.$$

An excellent reference for the proofs of the above proposition is the book of Vainberg [23].

3. Maximal Monotone Operators. An important property of convex functionals which have a gradient is the following:

Proposition 3.1. If $f(x)$ is convex then $F(x)$ satisfies the following inequality

$$\langle F(x) - F(y), x - y \rangle \geq 0, \quad x,y \in H.$$

Outline of proof. Consider the real-valued function

$$\phi(t) = f(y+th), \quad 0 \leq t \leq 1$$

where $h = x - y$. Because f is Gateaux differentiable and convex, the function $\phi(t)$ is convex and differentiable on $[0,1]$. Thus $\phi'(t)$ is increasing. But $\phi'(t) - \langle F(y+th), h \rangle$ and thus

$$\langle F(x) - F(y), x - y \rangle = \phi'(1) - \phi'(0) \geq 0.$$

The converse of the above is also true.

Proposition 3.2. If f is a functional on a Hilbert space H whose gradient F satisfies

$$\langle F(x) - F(y), x - y \rangle \geq 0, \quad x,y \in H,$$

then f is convex.

Proof. $\lambda f(x) + (1-\lambda)f(y) - f(\lambda x + (1-\lambda)y)$

$$= \lambda[f(x) - f(\lambda x + (1-\lambda)y)]$$
$$+ (1-\lambda)[f(y) - f(\lambda x + (1-\lambda)y)]$$
$$= \lambda\langle F[\lambda x + (1-\lambda)y + t_1(1-\lambda)(y-x)], (1-\lambda)(y-x)\rangle$$
$$+ (1-\lambda)\langle F[\lambda x + (1-\lambda)y + t_2\lambda(x-y)], \lambda(x-y)\rangle,$$

where $0 < t_1, t_2 < 1$.

The hypothesis on F now implies that

$$\lambda f(x) + (1-\lambda)y - f(\lambda x + (1-\lambda)y) \geq 0,$$

i.e., f is convex.

Combining the two propositions, we now obtain that f is convex if and only if F satisfies

$$\langle F(x) - F(y), x - y\rangle \geq 0,$$

$x,y \in H$. This leads to the following definition.

An operator $T: H \to H$ is said to be monotone if $\langle Tx - Ty, x - y\rangle \geq 0$ for all $x,y \in H$. If T is multivalued then the above definition is replaced by $\langle a - b, x - y\rangle \geq 0$ for all $x,y \in H$, $a \in Tx$, $b \in Ty$. The operator T is said to be maximal monotone if the graph of T is not properly contained in the graph of another monotone operator. It now follows from Section 2 that the minimum points of f are solutions of the operator equation $F(x) = 0$.

We now present some fundamental properties of maximal monotone operators.

Proposition 3.3. If $T: H \to H$ is single valued, continuous and monotone, then T is maximal monotone.

Proposition 3.4. Let $T: H \to H$ be monotone. Then T is maximal monotone if and only if the equation $(I+T)x = a$ is solvable for every $a \in H$.

Combining Propositions 3.3 and 3.4 we note that if f is a convex differentiable functional with a continuous gradient, then F is maximal monotone.

Proposition 3.5. Let $T: H \to H$ be maximal monotone and let $R > 0$ be such that $\langle Tu, u\rangle \geq 0$ for all $\|u\| \geq R$. Then the equation $Tu = 0$ has a solution.

The last proposition can thus be looked at as a sufficient condition for a convex function with a continuous gradient to have an extremum.

We now consider the case when f is not necessarily differentiable. The <u>subdifferential</u> of f at a point $x \in H$ is defined by

$$\partial f(x) = \{z \in H \mid f(u) - f(x) \geq \langle u - x, z \rangle \text{ for every } u \in \text{dom } f\}$$

The subdifferential of f is thus a not necessarily single valued operator. However if f has a gradient, then the subdifferential is single valued. The following properties of the subdifferential of f are immediate.

<u>Proposition 3.6</u>. Let $f: H \to R$ be convex and let $f \not\equiv +\infty$. Then ∂f is monotone.

<u>Proposition 3.7</u>. If $f: H \to R$ is convex and lower semicontinuous then f is weakly lower semicontinuous.

We now define the functional g by

$$g(x) = \frac{1}{2}\|x - a\|^2 + f(x), \quad \text{for a fixed}$$

$a \in H$. Then g is weakly lower semicontinuous. Further $g(x) \to \infty$ as $\|x\| \to \infty$. Hence by the results of Section 2, $g(x)$ attains its minimum at some $x_0 \in H$. Let $b = a - x_0$. Then it can be shown that $b \in \partial f(x_0)$, i.e., $a \in x_0 + \partial f(x_0)$. This implies that for any given $a \in H$, there exists $x_0 \in H$ such that $a \in x_0 + \partial f(x_0)$. By Proposition 3.4 we conclude that ∂f is maximal monotone.

For details of proofs of the above proposition and for further properties of monotone operators, we refer to Brezis [3].

4. <u>Algorithms for Approximate Solutions</u>.

In this section we present two algorithms for obtaining approximate solutions of nonlinear operator equations of the type $Fx = 0$ where $F: H \to H$ is a nonlinear maximal monotone operator. We will discuss the stochastic aspects of these algorithms in the following sections.

We first note that maximal monotonicity of F implies the same for cF where $c > 0$. Hence by the results quoted in the previous section $(I+cF)^{-1}y$ exists for each y. We now consider the sequence of elements $\{x_n\}$ defined by: for any arbitrary x_0 in H, let x_n be the unique solution of

$$Fx + \frac{x}{c_n} = g_n$$

where $c_n > 0$, $c_n \to \infty$ and $g_n \in H$. We then have:

<u>Proposition 4.1</u>. If $\|g_n\|$ and $c_n\|g_n\| \to 0$, then $\{x_n\}$ converges to a solution of $Tx = 0$ and the limit is the element of minimum norm in $T^{-1}(0)$.

Proof. The proof follows the lines of Dolph and Minty [5] and is related to the Tikhonov regularization method. Thus let $\bar{x}$ be an element of $F^{-1}(0)$. Then

$$Fx_n + \frac{x_n}{c_n} = g_n$$

and

$$F\bar{x} = 0.$$

Thus $\langle Fx_n - F\bar{x}, x_n - \bar{x}\rangle + \langle \frac{x_n}{c_n}, x_n - \bar{x}\rangle = \langle g_n, x_n - \bar{x}\rangle$. Using the monotonicity of F, we get

$$\|x_n\|^2 - \|x_n\| \, \|\bar{x}\| \leq c_n \|g_n\| \, \|x_n - \bar{x}\|.$$

From the hypothesis that $c_n \|g_n\| \to 0$, it follows that $\{x_n\}$ is bounded and let y be a weak limit of a subsequence of $\{x_n\}$ which will be denoted by $\{x_n\}$.

Now, $\langle Fx - Fx_n, x - x_n\rangle \geq 0$ and $Fx_n + \frac{x_n}{c_n} = g_n$ imply

$$\langle Fx, x - x_n\rangle + \frac{1}{c_n}\langle x_n, x - x_n\rangle - \langle g_n, x - x_n\rangle \geq 0.$$

Using the fact that $c_n \to \infty$ and $\|g_n\| \to 0$, in the limit we have

$$\langle Fx, x - y\rangle \geq 0$$

and, since F is maximal, $Fy = 0$.

Using similar arguments we can show that y is the element of minimum norm of $F^{-1}(0)$.

The above algorithm involves solving a nonlinear equation at each step. This is avoided in the following fashion. If F is of the type $cI + T$ where T is maximal monotone and $c > 0$, then there exists $x \in H$ such that $0 \in Fx = cx + Tx$. If T is Lipschitz continuous, we consider the scheme

$$x_{n+1} = x_n - \lambda(cx_n + Tx_n).$$

For λ sufficiently small it can be shown that x_n converges strongly to a zero of F. If T is not Lipschitz continuous, we can still consider the above scheme modified by

$$x_{n+1} = x_n - \lambda_n(cx_n + Tx_n).$$

This scheme may be seen in Bruck [4].

When F is not of the type $cI + T$ one could not always guarantee strong convergence of x_n. However we can resort once again to the Tikhonov regularization method and proceed as follows: apply the scheme above not to F but to $F + \phi_n I$, $\phi_n > 0$. We then let $\phi_n \to 0$ and choose ϕ_n and λ_n in a proper fashion to guarantee convergence of x_n where x_n is now given by

$$x_{n+1} = x_n - \lambda_n(Fx_n + \phi_n x_n).$$

We refer the reader to Bruck [4] for details.

We conclude this section with a detailed discussion of a particular case of the above algorithm in view of the results of Section 6. Thus let B be a bounded closed convex subset of a Hilbert space H and let M be a map of B into itself satisfying

$$\langle Mx - My, x - y\rangle < \gamma\|x - y\|^2, \quad \gamma \leq 1, \quad x,y \in B.$$

Further let M be Lipschitz continuous. Then the map $T = I - M$ satisfies

$$\langle Tx - Ty, x - y\rangle \geq (1-\gamma)\|x - y\|^2, \quad x,y \in B.$$

Also, for ε sufficiently small, $M_\varepsilon : B \to B$ defined by $M_\varepsilon x = (1-\varepsilon)x + \varepsilon M$ is a strict contraction and any solution of the equation $M x = x$ is also a solution of $Tx = 0$. This leads us to consider the sequence $\{x_n\}$ defined by

$$x_{n+1} = (1-a_n)x_n + a_n Mx_n = x_n - a_n(I-M)x_n = x_n - a_n Tx_n.$$

It can now be shown that this sequence converges strongly to the solution of the equation $Tx = 0$ if a_n satisfies the growth hypotheses that $\lim a_n = 0$ and $\Sigma a_n = \infty$.

We will utilize this algorithm in Section 6 to study the case of random errors. The hypotheses on T are rather strong and can be weakened. The details of these more general results will be published elsewhere.

5. The Stochastic Case.

As we have seen in the earlier sections the problem of minimizing a convex functional is equivalent to solving a nonlinear operator equation involving monotone operators. And in Section 4 we have noted some algorithms for finding approximate solutions of these nonlinear operator equations. But there are two situations that are of importance.

The first is the case when the algorithm involves errors. This situation arises as follows. For a convex nonlinear functional f the algorithm for finding approximations to the minima involves the gradient of f. However in a given situation we may not be able to compute this gradient but only observe it at a point. This observation naturally incurs an error which is a random variable. Thus the algorithm now reduces to

$$x_{n+1} = x_n - c_n(F(x_n)+z_n)$$

where F is the gradient of f and z_n is a sequence of random variables. We now have to establish the convergence of x_n to a minimum point in some probabilistic sense and also obtain estimates on the moments of the error generated by choosing an approximate solution.

This algorithm is very closely related to the Robbins-Munro procedure [19] of stochastic approximation. Briefly this may be described as follows. For each number x, Y_x is a random variable having a variance which is a bounded function of x; i.e., $E(Y_x - E(Y_x))^2 < \sigma^2 < \infty$. The regression curve $y = \phi(x) = E(Y_x)$ is presumed to be unknown but supposed to lie below the horizontal line $y = \alpha$ for $x < \theta$ and above it for $x > \theta$, where α is specified and θ is to be estimated. Let

$$x_{n+1} = x_n - a_n(Y_{x_n} - \alpha), \quad n = 1,2,...$$

where the a_n are specified numbers and assume that $E(x_1-\theta)^2 = v^2 < \infty$; we can then prove the convergence of x_n to θ under suitable hypotheses.

In the particular case when $\alpha = 0$ and $Y_x = \text{grad } f(x) + z(x)$ we see that the above stochastic approximation procedure reduces to our algorithm. In Section 6 we will study the convergence of our algorithm and obtain estimates on the moments of the random variables $\{x_n\}$.

As remarked above, the stochastic approximation algorithms may thus be considered as modifications of gradient methods for finding minima of functions. This leads to the question of considering the continuous time analogue of the deterministic case in the stochastic case. Continuous time stochastic approximation procedures have not been studied very extensively. We refer to [7, 8, 18, 20] for some of the results in this direction.

In the deterministic case the continuous time algorithm reduces to the equation

$$\dot{x}(t) = -a(t)F(x)$$

$$x(t_0) = x_0.$$

For the convergence, i.e., limit as $t \to \infty$ of $x(t)$ approaching x^*, one is referred to [9].

Analogously in the stochastic case we have the following equation:

$$\dot{x}(t,\omega) = -a(t)F(\omega,t,x(t))$$

where F is the gradient of f. Note that in this case f being a random functional implies that one would have to consider, as in the deterministic case, the analogous random operator equation $F(\omega,x) = 0$. The theory of random operators and random operator equations has been studied extensively in the recent years and we refer to [2, 13, 21] for details. Returning to the continuous time stochastic problem, one has to find sufficient conditions in order that

$$\rho\{x(t,\omega), S\} = 0 \quad \text{a.s.}$$

where S is the set of minima of f. We will discuss this problem in detail in Section 7.

In Section 4, as one of the algorithms for solving the minimization problem we discussed the Tikhonov regularization method. In Section 8 we consider the stochastic analogue of the Tikhonov regularization procedure and some of the recent results that have been obtained in this direction.

We conclude this section by remarking that for the rest of this paper (Ω,B,μ) is a complete probability space. For any separable Hilbert space X, a map T: $\Omega \times X \to X$ is said to be a random operator if the function $\langle T(\omega)x, y\rangle$ is a scalar-valued random variable for every $x,y \in X$. In other words, $T(\omega)$ is a random operator if and only if $T(\omega)x$ is a X-valued random variable for every $x \in X$. An operator equation of the form $T(\omega)x = y(\omega)$, where $y(\omega)$ is a given X-valued random variable, is said to be a random operator equation; and any X-valued random variable $x(\omega)$ which satisfies the condition $\mu\{\omega: T(\omega)x = y\} = 1$ is said to be a random solution of the equation $T(\omega)x = y(\omega)$. In recent papers [2, 14, 15, 16] we have discussed the question of existence of random solutions of random operator equations in detail.

Let B_0 be a σ-field contained in B. Then the conditional expectation of a X-valued random variable $x(\omega)$ given B_0 is defined as the random variable $E[x|B_0](\omega)$ and this random variable satisfies

(i) $E[x|B_0](\omega)$ is measurable with respect to B_0 and $E[x|B_0] \in L^1[\Omega,X]$;

(ii) $\int_A E[x|B_0](\omega)d\mu = \int_A x(\omega)d\mu$ for every $A \in B_0$.

An important tool made use of in the next section is the following result due to Driml and Hans [6]: If B_0 is a σ-field contained in B and x is a X-valued random variable which is measurable with respect to B_0, then for any X-valued random variable $y(\omega)$ we have $E(\langle x,y\rangle|B_0) = \langle x, E(y|B_0)\rangle$.

Finally we state the following result on convergence of a series of real valued random variables [24]:

The series of real valued random variables Σz_n converges a.s., if for a sequence of σ-fields $\{B_n\}$ such that $\{z_1, \ldots, z_{n-1}\}$ is B-measurable, $\Sigma E(z_n^2) < \infty$ and $\Sigma E(z_n|B_n)$ converges a.s.

6. Convergence of the Algorithm: Discrete Time. We consider the sequence of elements $\{x_n\}$ defined by

$$x_{n+1} = x_n - c_n[F(x_n) + z_n]$$

where $\{z_n\}$ is a collection of random variables, f is a convex functional on a real Hilbert space and F is its gradient. Before we proceed to discuss the convergence of this sequence of H-valued random variables, we recall that on f the functional to be minimized, we assume that

(i) f is bounded from below;

(ii) $\|F(x) - F(y)\| \leq L\|x - y\|$, $L > 0$, $x,y \in H$.

The justification for hypothesis (i) may be seen from the results quoted in Section 2. The Lipschitz continuity of F is assumed as per the discussions at the end of Section 4 and from these same discussions is motivated the requirement that

(iii) c_n is a sequence of positive numbers such that $\Sigma c_n^2 < \infty$ and $\Sigma c_n = \infty$.

We now make the following assumptions on z_n:

(iv) $z_n \in L^2(\Omega)$ and $\Sigma c_n^2 E[\|z_n\|^2] < \infty$;

(v) $E(z_n|x_n) = 0$.

Clearly $x_n(\omega)$ is a H-valued random variable. It now follows by induction that $F(x_n(\omega)) \in L^2(\Omega)$. We now observe that $E\{\langle F(x_n), z_n\rangle |x_n\} = \langle F(x_n), E(z_n|x_n)\rangle = 0$ as discussed in Section 5. By using the mean-value theorem and the Lipschitz continuity of F we see that

$$E(f(x_n)-f(x_{n+1})) \geq -\left(\frac{c_n^2L}{2}\right)E(\|z_n\|^2) + (c_n-c_n^2L/_2)E(\|F(x_n)\|^2).$$

Summing, we obtain letting $\gamma = \inf f(x)$,

$$E[f(x_0)] - \gamma \geq - \sum^n (c_i^2L/_2)E(\|z_i\|^2) + \sum^n (c_i-c_i^2L/_2)E(\|F(x_i)\|^2).$$

This implies $\sum^\infty c_nE(\|F(x_n)\|^2)$ converges. Thus for n,m sufficiently large

$$E(f(x_n)-f(x_{n+m})) \leq \sum_n^{m+n-1} c_iE(\|F(x_i)\|^2).$$

Thus $E(f(x_n))$ converges.

We now show that $E(f(x_n)) \to \gamma$. To this end we obtain an estimate on $E(\|x_n - y\|)^2$ and use the hypothesis $\Sigma c_n = \infty$ to obtain a contradiction that $\lim E(f(x_n)) - \gamma > 0$. Once again we utilize (v) in establishing $E(\langle F(x_n), z_n\rangle | x_n) = 0$.

We remark here that hypothesis (v) may be interpreted as follows: the average error committed when we know precisely the point at which the gradient is being measured is zero.

We can now show that if in addition f is strictly monotone, $x_n(\omega) \to x^*$ a.s., where x^* is the unique minimum point (because of the strict convexity of f). The proof follows lines similar to Venter [24]. For an arbitrary $r > 0$ we define

$$c_n = \{\omega: \|x_n(\omega) - c_nF(x_n(\omega)) - x^*\|^2 < r^2\}.$$

Further we define

$$\beta_n = \begin{cases} 0 & \omega \in C_n \\ \dfrac{2\langle x_n - c_nF(x_n) - x^*, c_n z_n\rangle}{\|x_n - x^*\|^2} & \omega \in C_n^c. \end{cases}$$

We then show that $\Sigma\beta_n \to 0$ a.s. This leads to $\limsup\|x_{n+1}(\omega) - x^*\|^2 \leq r^2$ and hence $x_n(\omega) \to x^*$ a.s.

We conclude finally by remarking that one can also obtain estimates on $\limsup E(\|x_n - x^*\|^2)$ as in the literature on stochastic approximation. The details of proofs of the above results may be seen in [17, 22]. Finally it must be remarked that one must obtain convergence of the algorithm by relaxing the hypothesis of Lipschitz continuity on F, as is known from the deterministic theory. In particular it can be shown that the Lipschitz continuity of F can be replaced by requiring suitable boundedness of the range of F.

If the basic problem is that of minimizing a nonlinear convex functional subject to constraints then one can view the problem as that of minimizing f over a closed convex subset C of H. We then modify the algorithm as follows. Let P be the projection map of H onto C; i.e., for $x \in H$, Px is that element of C such that $\|x - Px\| = \inf_{y \in C} \|x - y\|$. The sequence of elements x_n is then defined by

$$x_{n+1} = P[x_n - c_n(F(x_n)+z_n)]$$

Finally there is need to modify the above algorithm to the case when the observation of F is replaced by observation of an appropriate difference quotient, which would be the case when observation can be made only on $f(x)$.

7. <u>Continuous Time Stochastic Approximation</u>. The continuous time analogue of the algorithm discussed in the previous sections may be stated as follows. Thus consider the problem of finding the vector x^* which minimizes the functional $f(x)$ in the admissible set $C = \{x: \phi_i(x) \leq 0, \; i = 1,2,\ldots,m\}$. Here $f(x)$, $\phi_i(x)$, $i = 1,2,\ldots,m$ are continuously differentiable functionals, defined everywhere in R^n. Further let $f(x)$ and $\phi_i(x)$, $i = 1,2,\ldots,m$ be convex and let x^* be the unique minimum. Using the method of penalty functions we can transform the above constrained problem to the minimization (unconstrained) of the function

$$F(x,t) = f(x) + a(t)I(x)$$

where t is a parameter, $a(t)$ is a scalar function, $I(x)$ is a continuously differentiable convex functional defined appropriately with respect to C. Then we can show that any sequence of unconstrained minimum points of F converge to x^* [9].

If we now use a time varying penalty so that the parameter t is replaced by continuous time t, then proceeding analogously the continuous time algorithm can be written as

$$\frac{dx}{dt} = -\nabla F(x,t), \quad x(0) = x_0$$

where ∇F is the gradient of F and $\nabla F(x,t) = \nabla f(x) + a(t)\nabla I(x)$. It must be noted that on $a(t)$ the assumptions are that $a(t) > 0$ and $a(t) \to \infty$ monotonically as $t \to \infty$.

Choosing $a(t) = t$ and considering the stochastic analogue of the above we can write the general algorithm of continuous stochastic approximation as follows:

$$\dot{x}(t,\omega) = -\frac{1}{t} f(\omega,t,x(\omega,t))$$

$$x(0,\omega) = x_0(\omega)$$

where the f is not to be confused with the f at the beginning of this section. The problem is to find sufficient conditions on f in order that the solution (random) $x(t,\omega)$ of the above problem exists and converges a.s. to a point x^*, which is a minimum point of f.

We can also characterize x^* in the following manner: let $r(x)$ (called the regression function) be such that

$$\mu\{\omega: \lim_{t\to\infty} \frac{1}{t}\int_0^t f(\omega,\tau,x)d\tau = r(x) \quad \text{for every } x\} = 1.$$

Further let $S = \{x: r(x) = 0\}$. Then the basic problem of continuous time stochastic approximation is to find sufficient conditions on f in order that the solution $x(t,\omega)$ of the problem is such that

$$\mu\{\omega: \lim_{t\to\infty} \rho(x(t,\omega), S) = 0\} = 1.$$

The connection between the above discussion and the method of averaging is now fairly evident. We will outline it only briefly here. Consider the random differential equation

$$\dot{x}_\eta(t,\omega) = \eta f(x_\eta(t,\omega),\omega,t), \quad x_\eta(0) = x_0$$

We consider the averaged equation

$$\dot{y}_\eta(t) = \eta E(f), \quad y_\eta(0) = x_0.$$

The problem now is to find sufficient conditions on f in order that $y_\eta(t)$ be a good approximation to the random function $x_\eta(t,\omega)$ as $\eta \to 0$.

In the rest of this section we shall state two theorems which answer the above problems partially: The first is due to Hanš [11].

Theorem 7.1. Let T be a generalized stochastic process mapping the Cartesian product space $\Omega \times [0,\infty) \times X$ into the space X and almost surely continuous with respect to both the arguments $t \geq 0$ and $x \in X$ simultaneously. Let there exist an element $\hat{x} \in X$, a real-valued random variable c, and let the following assumptions be satisfied:

(i) $\mu\{\omega: \; c(\omega) < 1\} = 1;$

(ii) $\mu\{\omega: \; \lim_{t\to\infty} \| t^{-1} \int_0 T(\omega,s,\hat{x})ds - \hat{x}\| = 0\} = 1;$

(iii) $\mu\{\omega: \; \|T(\omega,t,x) - T(\omega,t,y)\| \le c(\omega)\|x - y\|\} = 1;$

for every $t \ge 0$ and every two elements x and y from X. Further, let $x_t(\cdot)$ be the solution of the random operator equation

$$\xi_0(\cdot) = T[\cdot,0,\xi_0(\cdot)];$$

$$\xi_t(\cdot) = t^{-1}\int_0^t T[\cdot,x,\xi_s(\cdot)]ds \quad \text{for} \quad t > 0.$$

Then x_t is for every $t \ge 0$ a generalized random variable and we have

$$\mu\{\omega: \; x_t(\omega) \text{ is continuous in } t\} = 1$$

and

$$\mu\{\omega: \; \lim \|x_t(\omega) - \hat{x}\| = 0\} = 1.$$

Hanš states that the contraction property of f is too severe. In [22] a relaxation of this is obtained but the results are still not very satisfactory.

In [7] Driml and Nedoma prove the following:

Theorem 7.2. Let $f(\omega,t,x)$ be a random function such that

(i) $\mu\{\omega: \; f(\omega,t,x)$ is continuous with respect to t and x simultaneously$\} = 1;$

(ii) for every t

$\mu\{\omega: \; x_1 < x_2$ implies $f(\omega,t,x_1) \le f(\omega,t,x_2)$, $x_1,x_2 \in (-\infty,\infty)\} = 1;$

(iii) there exists a function $r(x)$ such that

$\mu\{\omega: \; \lim_{t\to\infty} \frac{1}{t}\int_0^t f(\omega,\tau,x)d\tau = r(x)$ for every $x\} = 1.$

Then the solution of the random differential equation

$$\dot{x}(\omega,t) = -\frac{1}{t} f(\omega,t,x)$$

exists with probability one and further

$$\mu\{\omega: \; \lim_{t\to\infty} \rho\{x(\omega,t), S\} = 0\} = 1$$

where $S = \{x: \; r(x) = 0\}$ and ρ is the distance on the real line.

In [17] we have used Lyapunov function techniques and "continuation of solutions" results from differential equations in abstract spaces to obtain stronger results than the above in a more general setting. These techniques have been applied to dynamical systems in the setting of Itô differential equations.

8. Tikhonov Regularization: Stochastic Case. We now survey some of the results obtained for approximate solutions of random linear problems by using Tikhonov regularization. Let $Lu = f$ describe a system with output f (known) and unknown u. This problem may be ill-posed. A small error in measuring f may produce a large variation in u and thus give misleading information about u.

Suppose that L is a linear operator mapping a Hilbert space H_1 into a Hilbert space H_2 such that L does not have a bounded inverse. The problem $Lu = f$ is assumed to be an ill-posed problem in that there may be no solution or there may be many solutions. In practice, we may not know f exactly but only up to random errors in measurement. Then the ill-posed problem reduces to

$$Lu = g + z$$

where z is the random error.

It must be noted that even if L^{-1} exists and is bounded, the problem may be ill-posed if the condition number $\|L\| \, \|L^{-1}\|$ is much greater than 1. Note that a perturbation of δf produces a perturbation δu whereby the relative modification is $\left(\frac{\|\delta f\|}{\|f\|}\right)^{-1}\left(\frac{\|\delta u\|}{\|u\|}\right)$. It can now be shown that

$$\sup\left(\frac{\|\delta f\|}{\|f\|}\right)^{-1}\left(\frac{\|\delta u\|}{\|u\|}\right) = \|L\| \, \|L^{-1}\| .$$

And thus a large condition number would lead to significant changes in u with very little change in f.

Returning to the stochastic case, we now have two unknowns u and z. Here u and z will be sample functions drawn from random processes over H_1 and H_2 respectively. In [10] Franklin discusses the stochastic analogue of the Tikhonov regularization method as applied to the above situation and studies several examples.

In [12] we have presented how the Tikhonov regularization scheme is used in a modified form to handle nonlinear problems.

Also in Section 4, the same method has been applied to obtain approximate solutions for nonlinear operator equations by algorithms analogous to the Robbins-Munro procedure.

Thus the corresponding stochastic case for the nonlinear problem can now be handled and these results will be published elsewhere.

However we would like to mention the work of Bakushinskii and Apartsin [1] in considering problems of the type

$$Lu = f$$

where L is assumed to be positive and noninvertible. They generate an algorithm of the Robbins-Munro type and prove that the algorithm converges in a probabilistic sense to the normal solution of the above problem. In Section 4 we point out how in the case of non-linear but monotone L similar algorithms hold in the deterministic case. It remains to be seen whether these results are true in the stochastic case.

REFERENCES

1. Bakushinskii, A.B. and Apartsin, A.S. (1975) Methods of stochastic approximation type for solving linear incorrectly posed problems, Siberian Math. J., 16, 9-14.

2. Bharucha-Reid, A.T. (1972) Random Integral Equations, Academic Press.

3. Brézis, H. (1973) Opérateurs maximaux monotones, North Holland Publishing Co., Amsterdam.

4. Bruck, R.E., Jr. (1974) A strongly convergent iterative solution of $o \in U(x)$ for a maximal monotone operator U in a Hilbert space, J. Math. Anal. Appl., 48, 114-126.

5. Dolph, C.L. and Minty, G.J. (1964) On nonlinear integral equations of the Hammerstein type, Nonlinear Integral Equations, U. of Wisconsin Press, 99-154.

6. Driml, M. and Hans, O. (1960) Continuous stochastic approximation, Trans, Second Prague Conf. Infor. Theory, Prague, 113-122.

7. Driml, M. and Nedoma, J. (1960) Stochastic approximations for continuous random processes, Trans. Second Prague Conf. Infor. Theory, Prague, 145-158.

8. Dupač, V. (1965) A dynamic stochastic approximation method, Ann. Math. Stat., 36, 1695-1702.

9. Fiacco, A.V. and McCormick, G.P. (1968) Nonlinear programming: sequential unconstrained minimization techniques, John Wiley.

10. Franklin, J.N. (1970) Well-posed stochastic extensions of ill-posed linear problems, Jour. Math. Anal. Appl., 31, 682-716.

11. Hans, O. (1961) Random operator equations, Proc. Fourth Berkeley Symp. Math. Stat. Prob., 185-201.

12. Kannan, R. (1978) Tikhonov regularization and nonlinear problems at resonance - deterministic and random, Nonlinear Functional Analysis: Collected papers in houour of E. Rothe, Academic Press.

13. Kannan, R. (1977) Random operator equations, Proc. Int. Conf. Nonlinear Analysis in Gainesville, Academic Press, 113-138.

14. Kannan, R. Random correspondences and nonlinear equations, J. Multivariate Analysis (to appear).

15. Kannan, R. (1978) Sur une analogue du théoreme du point fixé de Kakutani, C. R. Acad. Sci., Paris, 287, 551-552.

16. Kannan, R. and Salehi, H. (1977) Random nonlinear equations and monotonic nonlinearities, J. Math. Anal. Appl. 57, 234-256.

17. Kannan, R. and Sutherland, P. Stochastic approximation and nonlinear optimization (to appear).

18. Krasulina, T.P. (1971) On stochastic approximation for random processes with continuous time, Theory Prob. Appl. 16, 674-682.

19. Robbins, H. and Monro, S. (1951) A stochastic approximation method, Ann. Math. Stat., 22, 400-407.

20. Sackrison, D.J. (1964) A continuous Kiefer-Wolfowitz procedure for random processes, Ann. Math. Stat., 35, 590-599.

21. Skorohod, A.V. (1975) Random operators in a Hilbert space, Proc. Third Japan-USSR Symp. Prob. Theory, Lecture notes in Math. #550, Springer Verlag, 562-591.

22. Sutherland, P. (1978) Probabilistic analysis of random nonlinear equations, Ph.D. thesis, U. of Texas at Arlington.

23. Vainberg, M.M. (1964) Variational methods for the study of nonlinear operators, Holden Day Co.

24. Venter, J.H. (1966) On Dvoretzky stochastic approximation theorems, Ann. Math. Stat., 37, 1534-1544.

Approximation Methods for the Minimum Average Cost per Unit Time Problem with a Diffusion Model*

Harold J. Kushner

Division of Applied Mathematics
Brown University, Providence, Rhode Island 02912

Abstract
1. Introduction
2. A Dynamic Programming Sufficient Condition for Optimality for (1), (2)
3. Bonded State Space Approximations
4. The Submartingale Problem of Strook and Varadhan in G
5. Discretization
6. Weak Convergence
7. Optimality of the Limit $\xi(\cdot)$
References

*This research was partially supported by the U.S. Air Force Office of Scientific Research under AF-AFOSR 76-3063, in part by the U.S. National Science Foundation under NSF-Eng73-03846-A-03, and in part by the U.S. Office of Naval Research under N0014-76-C-0279-P0002.

Abstract

Approximation methods for the minimum average cost per unit time problem with a controlled diffusion model is treated. In order to work with a bounded state space, we use the reflecting diffusion model of Strook and Varadhan, although other models can also be treated. The control problem is approximated by an average cost per unit time problem for a Markov chain, and weak convergence methods are used to show convergence of the minimum costs to that for the optimal diffusion. The procedure is quite natural and allows the approximation of many interesting functionals of the optimal process.

1. Introduction. In this paper, we develop an approximation and computational approach to a particularly difficult class of stocastic control problems. The computational problem leads to the approximation of the original process and optimization problem by an interesting and simpler sequence of processes and optimization problems, which yields much information on the original optimal process.

Let $w(\cdot)$ denote an R^r-valued Wiener process, let $\mathcal{U}$ denote a compact set and define the bounded and continuous functions $f(\cdot,\cdot)$: $R^r \times \mathcal{U} \to R^r$; $k(\cdot,\cdot)$: $R^r \times \mathcal{U} \to R$; $\sigma(\cdot)$: $R^r \to r \times r$ matrices. Let $x(\cdot)$ denote a non-anticipative solution to the Itô equation

$$dx = f(x,u)dt + \sigma(x)dw, \tag{1}$$

where $u(\cdot)$ is a non-anticipative (always with respect to $w(\cdot)$) $\mathcal{U}$-valued progressively measurable control function. For typographical simplicity we sometimes write x_s for $x(s)$, etc. Define $\gamma^u(\cdot)$ by

$$\gamma^u(x) = \overline{\lim_{t\to\infty}}\, E_x^u \frac{1}{t}\int_0^t k(x_s,u_s)ds, \tag{2}$$

where E_x^u denotes the expectation when $x_0 = x$ and control $u(\cdot)$ is used.

We are interested in finding good approximations to the infimum $\bar{\gamma}$ of $\gamma^u(x)$ over all controls $u(\cdot)$, and to the optimal control, and also other information concerning the optimal trajectory, in cases where $\gamma^u(x)$ does not depend on the initial state x. Furthermore, we want to be able to compute the approximation and obtain the additional information by using practical computational methods.

A number of difficulties stand in the way of a practical computation. First, the state space R^r of $x(\cdot)$ is unbounded and the control problem (1) - (2) will have to be modified so that the state space is bounded. This is a particularly ticklish point, since we want a modification which yields usable information concerning the original problem. In particular situations, a great deal of attention must be devoted to this. For definiteness, we use the bounded process defined in Section 4, although many others are possible. Next, we have not assumed very much about the system (1). If $\gamma^u(\cdot)$ actually depends on x, then very little is known about the problem. Fortunately, for many problems (perhaps the most important ones) we can restrict attention to $u(\cdot)$ which are stationary ($u(\cdot)$ is a stationary process), or to the stationary pure Markov case (where $u_t = u(x_t)$). Even then, the solution to (1) may not be unique. In practical problems, it is often demanded that the system have a certain robustness. Criteria such as (2) are of interest when the system is to operate over a long period of time, usually of uncertain duration and with an uncertain initial condition. It is usually desired that the control be stationary pure Markov, and that for the controls $u(\cdot)$ in the class which are to be considered there be an invariant measure μ^u, and the measures of $x(t)$ tend to μ^u as $t \to \infty$ for each $x = x_0$. In certain cases (e.g., Kushner [1]) one can restrict attention to such controls. In general, little is known about the continuous parameter problem, and many of the difficulties in the way of establishing convergence of a computational procedure are due to this. Also, it is usually hard to approximate problems over an infinite time interval, unless the approximation and limit processes are stationary. Furthermore, the ergodic subsets for each approximation may depend on the approximation. In any case, the procedures to be developed here are very natural, provide much information, and do give the desired results under broad conditions. We will later make an additional assumption on the system.

Our approach follows the ideas in Kushner [2], [3] and Kushner and DiMasi [4]. The problem (1), (2) is approximated by a control problem on a Markov chain (with approximation parameter h), and weak convergence methods are used to show that certain interpolations of the sequence of approximating chains converge weakly to an optimal process. The method yields a great deal of information on the optimal process; e.g., invariant measures and joint distributions.

A formal dynamic programming approach to the optimization of (1), (2) is given in Section 2, Section 3 argues for a "computational approximation" and a bounded state space. The actual form of the bounded state space model, the Strook-Varadhan model of a reflected diffusion [5], is discussed in Section 4. This model is used partly for the sake of specificity and partly because it allows us to illustrate some interesting features of the weak convergence and boundary time scaling. The actual discrete state model is developed in Section 5, and Sections 6 and 7 give the weak convergence results.

2. <u>A Dynamic Programming Sufficient Condition for Optimality for (1), (2).</u> Let $\mathscr{L}^u$ denote the differential generator of (1):

$$\mathscr{L}^u = \sum_i f_i(x,u) \frac{\partial}{\partial x_i} + \sum_{i,j} a_{ij}(x) \frac{\partial^2}{\partial x_i \partial x_j},$$

$$a(\cdot) = \sigma(\cdot)\sigma(\cdot)'/2.$$

When evaluating $\mathscr{L}^u F(\cdot)$ at t,ω, for a $C^2(R^r)$ function $F(\cdot)$, set $x = x_t$, $u = u_t$. Suppose that there is a $C^2(R^r)$ function $V(\cdot)$ and a constant $\bar{\gamma}$ such that

$$\inf_{\alpha \in \mathscr{U}} [\mathscr{L}^\alpha V(x) + k(x,\alpha) - \bar{\gamma}] = 0, \tag{3}$$

where $\mathscr{L}^\alpha$ is now treated as a parametrized operator. If there is a Borel function $\bar{u}(\cdot)$ on R^r such that $\alpha = \bar{u}(x)$ minimizes at x in (3) for each $x \in R^r$, and to which there corresponds a process (1) such that $E_x^u V(x_t)/t \to 0$, then

$$\bar{\gamma} = \lim_{t\to\infty} \frac{1}{t} E_x^{\bar{u}} \int_0^t k(x_s,\bar{u}_s)\,ds. \tag{4a}$$

If, in addition, $v(\cdot)$ is any $\mathscr{U}$-valued non-anticipative (ω,t) progressively measurable function (henceforth called a control) corresponding to which there is a solution to (1), and if $\frac{1}{t} E_x^v V(x_t) \to 0$, then

$$\bar{\gamma} \leq \underline{\lim}_{t\to\infty} \frac{1}{t} E_x^v \int_0^t k(x_s,v_s)\,ds, \tag{4b}$$

and $\bar{u}(\cdot)$ is optimal with respect to such $v(\cdot)$ in the sense that $\gamma^{\bar{u}} \leq \gamma^{v}$ for any x_0 either fixed or random. Under $\bar{u}(\cdot)$ or $v(\cdot)$, (1) is homogeneous, but there is not necessarily a unique invariant measure.

3. Bounded State Space Approximations. The approximation and computational method developed in [2] is roughly as follows. Let $u(\cdot)$ be fixed, and let it be a function only of the state x. We derive a family (parametrized by h) of Markov chains. For fixed $u(\cdot)$, the sequence of (suitable) continuous parameter interpolations of the chains converge weakly to the solution to (1), as $h \to 0$, under broad conditions. For each h, we have a controlled (indexed by $u(\cdot)$) family of Markov chains. Optimize, using the appropriate Markov chain version of (2), and obtain the minimum value function for each chain. As $h \to 0$, the sequence of minimum values converges to the infimum, over a large class of comparison controls, of the value function of the original problem. Also, many properties of the approximations converge to similar properties of the limiting optimal process.

Since our interest is in feasible computations, as well as in convergence, it is necessary that for each h the state space of the approximating chain be finite. This requirement necessitates revision of the original system (1). The following are among several possibilities that can be dealt with.

(i) The state space may be naturally bounded, in that there are bounded sets G_0, G_1 such that if $x_0 \in G_0$, then $x_t \in G_1$ for all t and all $u(\cdot)$.

(ii) If $x_0 \in G_0$, then the approximating Markov chain remains in G_1, for all h, under the optimizing controls.

(iii) Impulsive control terms ([2], Chapter 8) are added to the cost function, such that the state is guaranteed to be "impulsively" driven into G_0, if it ever leaves G_1.

(iv) A bounded set G can be introduced, such that x_t is not allowed to leave $\bar{G} = G + \partial G$. To guarantee this, a suitable boundary process is introduced on ∂G.

For concreteness in the development, a particular form of (iv) will be dealt with. We let G be a hyper-rectangle and suppose that x_t is reflected from ∂G. A hyper-rectangle is chosen only to simplify the specification of the approximation on the boundary. Any region for which a specification with the proper convergence properties exists can be chosen.

4. The Submartingale Problem of Strook and Varadhan [5] in G.
In order to assure ourselves that the reflection is well defined, assume

(A1) for each i, $a_{ii}(x)$ is strictly positive on the boundary planes of $\bar{G}$ which are parallel to $\{x: x_i = 0\}$, where $x_i = i^{th}$ component of x.

We introduce a boundary control and cost function. Let $\mathscr{U}_0$ be a compact set, and define the bounded continuous functions $\gamma(\cdot,\cdot): \partial G \times \mathscr{U}_0 \to R^r$; $k_0(\cdot,\cdot): \partial G \times \mathscr{U}_0 \to R$; $\rho(\cdot): \partial G \to [0,1]$. Let the vector $\gamma(x,\alpha)$ with origin x point strictly interior to G for each $x \in \partial G$ and $\alpha \in \mathscr{U}_0$. For $A \subset R^r$, set $I_A(x)$ = indicator of set $\{x: x \in A\}$, let $x(\cdot)$ denote the generic element of $C^r[0,\infty)$ (R^r-valued continuous functions on $[0,\infty)$) as well as the solution to (1). Hopefully, no confusion will arise. Define $C_G^r = C^r[0,\infty) \cap \{x(\cdot): x_t \in \bar{G}$, all $t < \infty\}$ and $\mathscr{G}_t$ = σ-algebra on C_G^r induced by the projections x_s, $s \leq t$. For this reflecting diffusion, admissible controls $u(\cdot)$ are $\mathscr{U}$-valued when the process state $x_t \in \partial G$, and are $\mathscr{U}_0$-valued when the process state $x_t \in \partial G$.* For $q(\cdot,\cdot) \in C^{2,1}(\bar{G} \times [0,\infty))$,** and admissible $u(\cdot)$, define the function $F_q^u(\cdot,\cdot)$ on $C_G^r[0,\infty)$ by

$$(5) \qquad F_q^u(x(\cdot),t) = q(x_t,t) - q(x_0,0) - \int_0^t [\frac{\partial}{\partial s} + \mathscr{L}^u] q(x_s,s) I_G(x_s) ds.$$

For the moment, let $u(\cdot)$ depend only on the current state x. Suppose that for some $y \in \bar{G}$, there is a measure P_y^u on C_G^r such $P_y^u\{x_0 = y\} = 1$ and for each $q(\cdot,\cdot)$ in $C^{2,1}(\bar{G} \times [0,\infty))$ for which $\rho(x)q_t(x,t) + \gamma'(x,u(x))q_x(x,t) \geq 0$ for all $x \in \partial G$, and all $t \geq 0$, the process $\{F_q^u(\cdot,t), \mathscr{B}_t, P_y^u\}$ is a submartingale. Then P_y^u is said to solve the submartingale problem for initial value y. If, in the above, the vector y can be replaced by a measure ν_0 on $\bar{G}$, and $P_{\nu_0}^u\{x_0 \in \Gamma\} = \nu_0(\Gamma)$ for each Borel set Γ, then $P_{\nu_0}^u$ is said to solve the submartingale problem for initial measure ν_0.

If $u(\cdot)$ depends only on the current state x, then the solution

*And u_t is $\mathscr{G}_t$ measurable.

**Where $C^{2,1}$ is the set of uniformly bounded continuous functions on $\bar{G} \times [0,\infty)$ whose derivatives, up to second order in x and first in t, are continuous and uniformly bounded.

to the submartingale problem gives the desired reflected diffusion, and $\gamma(x,u(x))$ is the average "direction of reflection" at $x \in \partial G$, and $\rho(x)$ is a scale factor which determines the relative time that $x(\cdot)$ spends on ∂G ([2], [3], [5]). Since $\rho(\cdot)$ only affects the time scale, and not the costs ([3], [2], Chapter 10), for our modeling purpose it is sufficient to set $\rho(x) \equiv 1$, which we will do.

Let P_y^u solve the submartingale problem. There is a non-decreasing scalar valued process $\rho(\cdot)$, which only increases when $x_t \in \partial G$, and is such that for the above $q(\cdot,\cdot)$

$$F_q^u(x(\cdot),t) - \int_0^t [q_s(x_s,s) + \gamma'(x_s,u_s)q_x(x_s,s)]d\mu_s \tag{6}$$

is a martingale (with respect to $\{P_y^u, \mathscr{B}_t\}$). Furthermore, there is a standard Wiener process* $w(\cdot)$ such that under P_y^u, $(x(\cdot),u(\cdot),\mu(\cdot))$ are non-anticipative with respect to $w(\cdot)$ and w.p.1.

$$\begin{aligned} x_t = y &+ \int_0^t f(x_s,u_s)I_G(x_s)ds + \int_0^t \sigma(x_s)I_G(x_s)dw_s \\ &+ \int_0^t I_{\partial G}(x_s)\gamma(x_s,u_s)d\mu_s. \end{aligned} \tag{7}$$

For the control problem, we may wish to deal with a larger class of (admissible) controls than the stationary pure Markov class. We can still speak of a solution to the submartingale problem, but then the measure P_y^u or $P_{\nu_0}^u$ must be defined on the appropriate σ-algebra on the product space of C_G^r and the path space for the control process. If this extended submartingale problem has a solution, then the non-decreasing process $\mu(\cdot)$ and Wiener process $w(\cdot)$ will still exist and (6), (7) hold.

<u>A modified control problem</u>. Suppose that there is a solution to the submartingale problem corresponding to admissible control $u(\cdot)$, and initial condition y. Define $\gamma^u(y)$ now by

$$\gamma^u(y) = \overline{\lim_{t\to\infty}} \frac{1}{t} E_y^u\{\int_0^t k(x_s,u_s)I_G(x_s)ds + \int_0^t k_0(x_s,u_s)I_{\partial G}(x_s)d\mu_s\}. \tag{8}$$

---

*To construct the Wiener process $w(\cdot)$, we may have to augment the probability space by adding an independent Wiener process.

Since $\rho = 1$, we can set $\mu_s = s$. The formal dynamic programming equation (3) is replaced by

$$\inf_{\alpha \varepsilon \mathscr{U}} [\mathscr{L}^{\alpha} V(x) + k(x,\alpha) - \bar{\gamma}] = 0, \quad x \varepsilon G,$$

(9)

$$\inf_{\alpha \varepsilon \mathscr{U}_0} [V_x'(x)\gamma(x,\alpha) + k_0(x,\alpha) - \bar{\gamma}] = 0, \quad x \varepsilon \partial G,$$

where $V(\cdot)$ is now assumed to be bounded. If there is a solution to the submartingale problem corresponding to admissible control $v(\cdot)$ and initial condition y, and also a smooth function $V(\cdot)$ and constant $\bar{\gamma}$ solving (9), then

$$\bar{\gamma} \le \gamma^{v}(y). \tag{10}$$

If there is a Borel admissible control $\bar{u}(\cdot)$ which attains the infimum in (9), and for which the submartingale problem has a solution for each initial condition x, then $\bar{\gamma} = \bar{\gamma}^{u}(y)$ and $\bar{u}(\cdot)$ is optimal. <u>We emphasize that although</u> (9) <u>will serve as the basis of our approximation, it need not have a solution of any sort for our method to be valued</u>.

5. <u>Discretization</u>. There are a number of techniques for getting an approximating sequence of Markov chain control problems with the correct convergence properties. We use the method in [2] mainly because it is relatively straightforward, fairly well understood and we can refer to existing results. The method is based on a finite difference approximation with difference interval h. A particular (but natural) finite difference approximation to (9) is used. It makes no difference whether or not (9) has a smooth solution, for the finite difference approximation is <u>not</u> used to solve (9). After a suitable rearrangement, the coefficients of certain terms in the finite difference approximation will be transition probabilities for an approximating controlled Markov chain. This is the <u>only</u> use to which (9) will be put. The method gives us an approximating chain simply and automatically. A detailed outline of the method and of some of the convergence properties will be given, but many of the details which can be found in the basic references [2], [3], [4] will be omitted.

Let e_i = unit vector in i^{th} coordinate direction, and assume for convenience that each side of G is an integral multiple of h.

Let G_h denote the finite difference grid on G, and set $\partial G_h = \overline{G}_h - G_h$, where $\overline{G}_h$ is the finite difference grid on $\overline{G}$. Now, let us discretize (9). On ∂G, use the approximation

$$\text{(11)} \qquad \begin{aligned} V_{x_i}(x) &\to [V(x+e_i h) - V(x)]/h, \quad \text{if } \gamma_i(x,\alpha) \geq 0 \\ V_{x_i}(x) &\to [V(x) - V(x-e_i h)]/h, \quad \text{if } \gamma_i(x,\alpha) < 0. \end{aligned}$$

In G, use the approximation

$$\text{(12)} \qquad \begin{aligned} V_{x_i}(x) &\to [V(x+e_i h) - V(x)]/h, \quad \text{if } f_i(x,\alpha) \geq 0 \\ V_{x_i}(x) &\to [V(x) - V(x-e_i h)]/h, \quad \text{if } f_i(x,\alpha) < 0 \\ V_{x_i x_i}(x) &\to [V(x+e_i h) + V(x-e_i h) - 2V(x)]/h^2. \end{aligned}$$

The approximations for $V_{x_i x_j}(x)$, $i \neq j$, are long, and the reader is referred to [2], Chapter 6.2 for one set of possibilities. Simply to avoid writing these down here, we suppose that $\sigma(x)\sigma'(x)$ is diagonal. This assumption is not required by anything except our current laziness. It does not affect the outcome, only the precise form of the functions $Q_h(\cdot,\cdot)$ and $p^h(\cdot,\cdot)$ introduced below.

Define $Q_h(x,\cdot), \Delta t^h(x)$ and $\overline{Q}_h(x)$ by

$$\begin{aligned} Q_h(x,\alpha) &= h \sum_i |f_i(x,\alpha)| + \sum_i \sigma_i^2(x), && x \in G_h, \\ Q_h(x,\alpha) &= \sum_i |\gamma_i(x,\alpha)| , && x \in \partial G_h \\ \overline{Q}_h(x) &= \sup_\alpha Q_h(x,\alpha), \end{aligned}$$

(where α ranges over the appropriate set $\mathscr{U}$ or $\mathscr{U}_0$),

$$\begin{aligned} \Delta t^h(x) &= h/\overline{Q}_h(x) \quad \text{on } \partial G_h, \\ &= h^2/\overline{Q}_h(x) \quad \text{on } G_h. \end{aligned}$$

Approximating the derivatives in (9) by (11)-(12) and rearranging terms yields the following equation, where $\overline{\gamma}^h$ and $V^h(\cdot)$ are used to denote the solution to the discretized equation and we use the definitions $g^+(x) = \max[g(x),0]$ and $g^-(x) = \max[0,-g(x)]$.

$$(13)\quad h^2\bar{\gamma}^h = \inf_{\alpha\in\mathscr{U}} [-Q_h(x,\alpha)V^h(x) + \sum_{i,\pm} V^h(x\pm e_i h)(hf_i^{\pm}(x,\alpha) + \sigma_i^2(x)/2) + h^2 k(x,\alpha)],\quad x \in G_h,$$

$$h\bar{\gamma}^h = \inf_{\alpha\in\mathscr{U}_0} [-Q_h(x,\alpha)V^h(x) + \sum_{i,\pm} V^h(x\pm e_i h)\gamma_i^{\pm}(x,\alpha) + hk_0(x,\alpha)],\quad x \in \partial G_h.$$

Define $p^h(x,x\pm e_i h|\alpha) = (\text{coefficient of } V^h(x\pm e_i h))/\bar{Q}_h(x)$, $p^h(x,x|\alpha) = [\bar{Q}_h(x) - Q_h(x,\alpha)]/\bar{Q}_h(x)$. Divide (13) through by $\bar{Q}_h(x)$ and rearrange to get

$$(14)\quad V^h(x) + \bar{\gamma}^h \Delta t^h(x) = \inf_{\alpha\in\mathscr{U}} [\sum_{i,\pm} V^h(x\pm e_i h)p^h(x,x\pm e_i h|\alpha) + V^h(x)p^h(x,x|\alpha) + k(x,\alpha)\Delta t^h(x)],\quad x \in G_h,$$

and similarly for x in ∂G_h, where $\mathscr{U}$ and k are replaced by $\mathscr{U}_0$ and k_0, resp. Define $p^h(x,y|\alpha) = 0$ for all x,y other than $y = x$ or $y = x \pm e_i h$ for some i. Then $\{p^h(x,y|\alpha),\ x,\ y \in \bar{G}_h\}$ is a transition probability for a controlled Markov chain. Let $\{\xi_n^h\}$ denote the random variables of the chain, and define $\mathscr{U}(x) = \mathscr{U}$ in G, and $\mathscr{U}(x) = \mathscr{U}_0$ on ∂G, and redefine $k(x,\alpha)$ to equal $k_0(x,\alpha)$ for $x \in \partial G$. Then (14) can be rewritten in the form

$$(15)\quad V^h(x) + \bar{\gamma}^h \Delta t^h(x) = \inf_{\alpha\in\mathscr{U}(x)} [E_x^{\alpha} V^h(\xi_1^h) + k(x,\alpha)\Delta t^h(x)],\quad x \in \bar{G}_h.$$

In (13)-(15), we supposed that $\bar{\gamma}^h$ is a constant. This is almost equivalent to the assumption that there is only one recurrence class for the chain under the optimal control. If there is more than one recurrence class, the numerical problem is harder. Let us henceforth assume

(A2) For each small h and under each stationary pure Markov control, there is only one recurrence class.

This assumption seems to hold in very many cases of practical interest. It can be dispensed with, but then the problem of actually solving (13)-(15) is much harder. Under (A2), (15) can be solved by either Howard's iteration in policy space for semi-Markov processes, or by a version of the backward iteration method for the average cost per unit time problem (see, e.g., Schweitzer and Federgruen [8], but adapted to a semi-Markov process model). There is an optimal stationary pure Markov control $u^h(\cdot)$ for all small h, it is the minimizer in (15), and it is optimal with respect to all controls for the discrete problem. The "semi-Markov" point will be returned to below. The optimal solution is given in the first line of (19).

Discussion of (14). For $y \in G_h$, we have for any stationary pure Markov control $u(\cdot)$

(16a)
$$E_y^u[\xi_{n+1}^h - \xi_n^h | \xi_n^h = y, \ u(\cdot) \ \text{used}] = f(y,u(y))\Delta t^h(y),$$

$$\mathrm{cov}_y^u[\xi_{n+1}^h - \xi_n^h | \xi_n^h = y, \ u(\cdot) \ \text{used}] = \sigma(y)\sigma'(y)\Delta t^h(y) + o(\Delta t^h(y)), \quad x \in G_h.$$

For $y \in \partial G_h$,

(16b)
$$E_y^u[\xi_{n+1}^h - \xi_n^h | \xi_n^h = y, \ u(\cdot) \ \text{used}] = \gamma(y,u(y))\Delta t^h(y),$$

$$\mathrm{cov}_y^u[\xi_{n+1}^h - \xi_n^h | \xi_n^h = y, \ u(\cdot) \ \text{used}] = o(\Delta t^h(y)).$$

These "infinitesimal" properties (derived in [2], [3]), together with (15), suggest a close relation between the controlled chain, and the controlled reflected diffusion.

These relations are brought out quite clearly when the chain is suitably interpolated into a continuous parameter process, and (15), (16) suggest several useful interpolations. First, we note that solving (15) is the only computation that need be done. Equation (15) is not quite the dynamic programming equation for the average cost per unit time for the controlled chain $\{\xi_n^h\}$, since $\bar{\gamma}^h$ has a state dependent coefficient $\Delta t^h(\cdot)$. However, it is the dynamic programming equation for a semi-Markov process or, equivalently for the types of continuous parameter interpolations which are discussed below.

Let π^h denote the invariant measure which corresponds to the optimal control. Henceforth, unless otherwise mentioned, $\{\xi_n^h\}$ refers to the optimal chain, with initial measure π^h.

We now choose an interpolation method and show that the sequence of interpolated processes converges weakly to a solution to the submartingale problem corresponding to some admissible control

$u(\cdot)$, and that this solution is an optimal one, with cost rate $\bar{\gamma} = \lim_{h\to 0} \bar{\gamma}^h$.

Either of the following two piecewise constant interpolations will serve our purpose.

Interpolation 1. Define $\Delta t^h(\xi_i^h) = \Delta t_i^h$, $t_n^h = \sum_{i=0}^{n-1} \Delta t_i^h$. Define the semi-Markov process $\xi^h(\cdot)$ by $\xi^h(t) = \xi_n^h$ on $[t_n^h, t_{n+1}^h)$. This interpolation was used in [2], [3].

Interpolation 2. Let $\xi^h(\cdot)$ denote the Markov jump process on $\bar{G}_h$ defined by:

If $\xi^h(t) = y$, then the average additional time spent in state y before a jump is $\Delta t^h(y)$, and $P\{\text{next state} = y' \mid \text{current state} = y\} = p^h(y,y'|u^h(y))$. There is a slight ambiguity here since it is possible that $p^h(y,y|u^h(y)) > 0$. But, this should cause no confusion - for it simply means that there is a jump of "zero" magnitude. The average interjump times can be normalized to avoid this, but it hardly seems worthwhile. Note that

$$P\{\text{jump in } (t,t+\Delta] | \xi^h(t) = y\} = (\Delta/\Delta t^h(y)) + o(\Delta).$$

This interpolation is developed in Section 8 of [4].

Neither interpolation is always preferable to the other. Interpolation 2 could have been used in references [2], [3], but there did not seem to be a need for it then. There are advantages to having an interpolation which is a continuous parameter Markov chain in that certain concepts (such as stationarity) have a clearer meaning; on the other hand it is sometimes preferable to work with interpolation times that are deterministic functions of the current state, since then there are fewer random variables to worry about. The limiting processes (see Sections 6 and 7) are the same for both interpolations. In Case 2, the average sojourn time in a state y (before the next jump, whether of zero value or not) is $\Delta t^h(y)$, precisely the interpolation interval for Case 1. In both cases, the time spent at a state y on the boundary ($O(h)$, per sojourn) is greater than time spent at a state y in G_h ($O(h^2)$ per sojourn, unless there is the complete degeneracy $\sigma(y) = 0$). This property is a consequence of our definition of $\Delta t^h(y)$ for $y \in \partial G_h$

(to correspond to $\rho(y) \equiv 1$).

For either Interpolation 1 or 2,

$$\bar{\gamma}^h = \lim_{t\to\infty} E_x^h \int_0^t k(\xi_s^h, u_s^h)\,ds/t, \tag{17}$$

where $u_s^h = u^h(\xi_s^h)$, and E_x^h indicates that u^h is used. The invariant measure for either interpolation is μ^h, where

$$\mu^h(y) = \Delta t^h(y)\pi^h(y)/\sum_z \Delta t^h(z)\pi^h(z) \tag{18a}$$

Also,

$$\bar{\gamma}^h = \sum_y \mu^h(y)k(y,u^h(y)). \tag{18b}$$

Equations (17) and (18) are not hard to verify. For example, (18) follows from the ergodic theorems for Markov chains (see Chung [6], Section 1.15, Theorems 1, 2, 3; see also [2], Chapter 6.8, for similar calculations). It can also be obtained by direct verification of the Kolmogorov equation using the invariance of $\pi^h(\cdot)$ for the discrete parameter chain. To get (17) write u_i^h for $u^h(\xi_i^h)$ and use (15) and the same ergodic theorems to get

$$\begin{aligned} \bar{\gamma}^h &= \lim_{n\to\infty} [E_x^h \sum_{i=0}^{n-1} k(\xi_i^h,u_i^h)\Delta t_i^h / E_x^h \sum_{i=0}^{n-1} \Delta t_i^h] \\ &= \underset{(\text{w.p.}1)}{\lim_n} [\sum_{i=0}^{n-1} k(\xi_i^h,u_i^h)\Delta t_i^h / \sum_{i=0}^{n-1} \Delta t_i^h] \\ &= \underset{(\text{w.p.}1)}{\lim_n} \int_0^{t_n^h} k(\xi_s^h,u_s^h)\,ds/t_n^h = \underset{t\to\infty}{\underset{(\text{w.p.}1)}{\lim}} \int_0^t k(\xi_s^h,u_s^h)\,ds/t \\ &= \lim_{t\to\infty} \int_0^t E_x^h k(\xi_s^h,u_s^h)\,ds/t. \end{aligned} \tag{19}$$

Similarly, the first limit in (19) equals

$$\begin{aligned} \bar{\gamma}^h &= \sum_y \pi^h(y)k(y,u^h(y))\Delta t^h(y) / \sum_y \pi^h(y)\Delta t^h(y) \\ &= \sum_y \mu^h(y)k(y,u^h(y)). \end{aligned} \tag{20}$$

Let $v(\cdot)$ denote a stationary pure Markov control. Then (15) implies that (here $\Delta t_i^h, \xi_i^h$ now refer to the variables under control $v(\cdot)$) for any x

$$\bar{\gamma}^h \leq \lim_{n\to\infty} \frac{E_x^v \sum_{i=0}^{n-1} \Delta t_i^h k(\xi_i^h, v(\xi_i^h))}{E_x^v \sum_{i=0}^{n-1} \Delta t_i^h} \equiv \gamma^{v,h}. \tag{21}$$

The proof of optimality of $u^h(\cdot)$ with respect to any control which is not necessarily stationary pure Markov can be based on a method of Ross [7] and is omitted.

6. Weak Convergence. We will work with Interpolation 2, since it is a strictly stationary process. The method will be outlined, but the proofs will be usually referred to when already available elsewhere. So far, we have a sequence of stationary pure Markov controls $\{u^h(\cdot)\}$, corresponding stationary continuous parameter Markov chains $\{\xi^h(\cdot)\}$, invariant measures $\{\mu^h\}$, and minimum costs $\{\bar{\gamma}^h\}$, where

$$\bar{\gamma}^h = \sum_{y\in\bar{G}_h} \mu^h(y)k(y,u^h(y)) = \sum_{y\in G_h} \mu^h(y)k(y,u^h(y))$$

$$+ \sum_{y\in\partial G_h} \mu^h(y)k_0(y,u^h(y)),$$

and

$$\bar{\gamma}^h t = E^h\left[\int_0^t k(\xi_s^h,u_s^h)I_G(\xi_s^h)ds + \int_0^t k_0(\xi_s^h,u_s^h)I_{\partial G}(\xi_s^h)ds\right], \tag{22}$$

where E^h denotes the expectation under initial measure μ^h, and we use $u_s^h = u^h(\xi_s^h)$. We often write $\xi^h(s)$ as ξ_s^h, etc., for typographical simplicity.

We obviously can write

$$\xi_s^h = \xi_0^h + \int_0^t I_G(\xi_s^h)f(\xi_s^h,u_s^h)ds \tag{23}$$

$$+ \int_0^t I_{\partial G}(\xi_s^h)\gamma(\xi_s^h,u_s^h)ds + B^h(t) + B_0^h(t),$$

where

$$B^h(t) = \int_0^t I_G(\xi_s^h)[d\xi^h(s) - f(\xi_s^h, u_s^h)ds],$$

$$B_0^h(t) = \int_0^t I_{\partial G}(\xi_s^h)[d\xi_s^h - \gamma(\xi_s^h, u_s^h)ds].$$

Denote the two integrals in (22) by $K^h(t)$ and $K_0^h(t)$, resp., and the first two integrals on the right side of (23) by $Q^h(t)$ and $Q_0^h(t)$, resp. Let $D^m[0,\infty)$ denote the space of R^m valued functions on $[0,\infty)$, continuous on the right and with left-hand limits (Billingsley [9], Lindvall [10], Kushner [2], Chapter 2), endowed with the Skorokhod topology. If a measure ν_n induces a process $X^n(\cdot)$ with paths in $D^m[0,\infty)$ w.p.1 and $\{\nu_n\}$ is tight, we abuse terminology and say that $\{X^n(\cdot)\}$ is tight. If $\{\nu_n\}$ converges weakly to a measure ν and ν induces a process $X(\cdot)$ with paths in $D^m[0,\infty)$ w.p.1, we say that $\{X^n(\cdot)\}$ converges weakly to $X(\cdot)$. We occasionally use Skorokhod imbedding ([11], Theorem 3.1.1, or [2], Chapter 2), which says that if $X^n(\cdot) \to X(\cdot)$ weakly in $D^m[0,\infty)$, then there are processes $\tilde{X}(\cdot), \tilde{X}^n(\cdot)$ with paths in $D^m[0,\infty)$ and which induce the same measures on $D^m[0,\infty)$ as do $X(\cdot), X^n(\cdot)$, resp., and are such that $\tilde{X}^n(\cdot) \to \tilde{X}(\cdot)$ w.p.1 in the Skorokhod topology. Since all our limit processes will be continuous w.p.1, this implies that $\tilde{X}^n(t) \to \tilde{X}(t)$, uniformly on bounded intervals. Also, we omit the tilde ~ notation. The following theorem follows from the results in [4], Section 8.

<u>Theorem 1.</u>* $\{\xi^h(\cdot), K^h(\cdot), K_0^h(\cdot), B^h(\cdot), B_0^h(\cdot), Q^h(\cdot), Q_0^h(\cdot)\} \equiv \{\Phi(\cdot)\}\}$ <u>is tight on</u> $D^{5r+2}[0,\infty)$, <u>and all limits have continuous paths</u> w.p.1.

We will next characterize the limits of $\{B^h(\cdot), B_0^h(\cdot)\}$.

Let us choose a weakly convergent subsequence, also indexed by h, and henceforth fixed. The subsequent results will not depend upon the selected subsequence. Denote the limit by $\xi(\cdot)$, $K(\cdot)$, $K_0(\cdot)$, $B(\cdot)$, $B_0(\cdot)$, $Q(\cdot)$, $Q_0(\cdot)$. By construction, $B^h(t)$ and

---

*Theorem 1 does not require A1 or A2 and holds whether the initial conditions are random or not. It needs only the boundedness and continuity of f, σ, k, k_0 and γ. Also, u^h can be replaced by any pure Markov control.

$B_0^h(\cdot)$ are martingales (with respect to the σ-algebras B_t^h induced by ξ_s^h, $s \le t$) and an easy calculation yields that

$$E \sup_{t \le T} |B_0^h(t)|^2 \le \text{constant} \cdot hT.$$

Thus $B_0(\cdot)$ is the zero process.

The quadratic variation of $B^h(\cdot)$ is

$$\int_0^t \Sigma^h(\xi_s^h) I_G(\xi_s^h) ds,$$

where $\Sigma^h(x)$ is such that it converges to $\sigma(x)\sigma'(x)$ as $h \to 0$, uniformly in x, and $\sup_h E|B^h(t)|^4 < \infty$ for each $t > 0$. Then $\{|B^h(t)|^2\}$ is uniformly integrable for each t. Let $\mathscr{B}_t$ denote the σ-algebra induced by $\{\xi_s, B(s), K(s), K_0(s), Q(s), Q(s),\ s \le t\}$. Let N_ε denote an ε neighborhood of ∂G. In [3], Lemma 1, it is shown that for each real $T > 0$ there is a constant K such that, for Interpolation 1 and small $\varepsilon > 0$

$$E_x^u \int_0^T I_{N_\varepsilon}(\xi_s^h) I_G(\xi_s^h) ds \le K_T \varepsilon, \tag{24}$$

uniformly in u, h (although u did not appear in the derivation, only an upper bound to the values of the drift function f was used in the derivation). The result (24) depends only on the fact that the component of the diffusion term $\sigma(x)dw$ orthogonal to the boundary is uniformly non-degenerate on ∂G; i.e., on (A1). Estimate (24) also holds for Interpolation 2, and is crucial for the rest of the development. It says that neither the approximations nor the limit can "linger" near (but not on) the boundary. In particular, it implies that the probability is zero that over some subinterval of $[0,T]$ the paths for the approximations will be in $N_\varepsilon \cap G$ and the limit will be on ∂G.

Theorem 2. Assume A1. $\{B(t), \mathscr{B}_t\}$ is a continuous martingale with quadratic covariation $\int_0^t I_G(\xi_s)\sigma(\xi_s)\sigma'(\xi_s)ds$.

Proof. The proof, using (24), follows similar calculations in [2], [3], [4]. Let $q^h(t)$ represent any of the vectors in $\Phi^h(\cdot)$ (see Theorem 1), let n denote an arbitrary integer, t_i, $i \le n$, numbers less than or equal to t, let $s > 0$ and let $g(\cdot)$ denote a continuous real valued function. By weak convergence, Skorokhod imbedding and the uniform integrability of $\{|B^h(t)|\}$ for each t, the result (martingale property of $B^h(\cdot)$)

$$E^h g(q^h(t_i), i \le n)[B^h(t+s) - B^h(t)] = 0$$

implies

$$Eg(q(t_i), i \le n)[B(t+s) - B(t)] = 0.$$

Also, the result

$$E^h g(q^h(t_i), i \le n)[(B^h(t+s) - B^h(t))(B^h(t+s) - B^h(t))' - \int_0^t I_G(\xi_s^h)\Sigma^h(\xi_s^h)ds] = 0$$

together with the weak convergence, Skorokhod imbedding and uniform* integrability of $\{|B^h(t)|^2\}$ and (24) implies that

$$Eg(q(t_i), i \le n)[(B(t+s) - B(t))(B(t+s) - B(t))' - \int_0^t I_G(\xi_s)\sigma(\xi_s)\sigma'(\xi_s))ds] = 0.$$

The arbitrariness of $g(\cdot)$, t, $t + s$, t_i, $i \le n$, and n imply the theorem. Q.E.D.

We next need a representation for $Q(\cdot)$, $Q_0(\cdot)$, $K(\cdot)$ and $K_0(\cdot)$. It is easy to see that all these functions are absolutely continuous with respect to Lebesgue measure. Thus, there are measurable (ω,t) functions $\overline{q}(\cdot)$, $\overline{q}_0(\cdot)$, $\overline{k}(\cdot)$ and $\overline{k}_0(\cdot)$ such that, for almost all ω,t,

$$Q(t) = \int_0^t \overline{q}(s)ds, \qquad Q_0(t) = \int_0^t \overline{q}_0(s)ds$$

$$K(t) = \int_0^t \overline{k}(s)ds, \qquad K_0(t) = \int_0^t \overline{k}_0(s)ds.$$

---

*Actually, uniform integrability of $\{|B^h(t)|^2\}$ (implied by $\sup_h E^h|B^h(t)|^4 < \infty$) is not needed. Since $B(\cdot)$ is a square integrable continuous martingale, its quadratic variation can be obtained by a "localization" of the argument.

We can now proceed in two ways, either working with generalized random controls or by imposing a convexity condition and using an implicit function theorem. We take the latter (and easier) approach.

Theorem 3.* Assume A1 and A2. Let f,k,k_0,γ,σ be continuous and let the sets $\{f(x,\alpha),\ k(x,\alpha),\ \alpha \in \mathscr{U}\} \equiv g(x,\mathscr{U})$ and $\{\gamma(x,\alpha),\ k_0(x,\alpha),\ \alpha \in \mathscr{U}_0\} \equiv g_0(x,\mathscr{U}_0)$ be convex for each x. Then there is a control $\bar{u}(\cdot)^+$ with values $\bar{u}_s$ in $\mathscr{U}$ when $\xi_s \in G$ and in $\mathscr{U}_0$ when $\xi_s \in \partial G$ and such that, for almost all ω,t,

$$\bar{f}(t) = f(\xi_t,\bar{u}_t)I_G(\xi_t)$$

$$\bar{f}_0(t) = \gamma(\xi_t,\bar{u}_t)I_{\partial G}(\xi_t)$$

$$\bar{k}(t) = k(\xi_t,\bar{u}_t)I_G(\xi_t)$$

$$\bar{k}_0(t) = k_0(\xi_t,\bar{u}_t)I_{\partial G}(\xi_t).$$

Proof. Define $\bar{g}(t) = (\bar{f}(t),\bar{k}(t))$ and $\bar{g}_0(t) = (\bar{f}_0(t),\bar{k}_0(t))$. The proof uses the basic estimate (24) and the method of [2], pp. 182-183. By (24) and [2], pp. 182-183, for almost all ω,t

$$\bar{g}(t) \in g(\xi_t,\mathscr{U})I_G(\xi_t)$$

$$\bar{g}_0(t) \in g_0(\xi_t,\mathscr{U}_0)I_{\partial G}(\xi_t),$$

from which the result follows by the McShane-Warfield implicit function theorem as in [2], Theorem 9.2.2. Q.E.D.

Summing up the results of Theorems 1 to 3, we get the representation (under A1 and A2)

$$(25)\quad \xi_t = \xi_0 + \int_0^t I_G(\xi_s)f(\xi_s,\bar{u}_s)ds + \int_0^t I_{\partial G}(\xi_s)\gamma(\xi_s,\bar{u}_s)ds + B(t),$$

where $B(t)$ is a continuous martingale with quadratic variation

$$\int_0^t I_G(\xi_s)\sigma(\xi_s)\sigma'(\xi_s)ds.$$

*This control is also non-anticipative with respect to the $w(\cdot)$ introduced below (25).

Also, there is a Wiener process $w(\cdot)$, with respect to which all the other processes in (25) are non-anticipative and such that $B(t) = \int_0^t I_G(\xi_s)\sigma(\xi_s))dw(s)$. Obviously, by the weak convergence, ξ_t is in $\bar{G}$ for all t. Let $\mathscr{L}^{\bar{u}}$ denote the differential generator associated with (25) in G. By a slight modification of the argument associated with (40) and (41) in [3], we can show that $\xi(\cdot)$ solves the sub-martingale problem.

Furthermore, $\xi(\cdot)$ is a stationary process. Let its invariant measure be denoted by μ (which is the weak limit of $\{\mu^h\}$), and let $\bar{\gamma} = \lim_h \bar{\gamma}^h$. Then the distribution of ξ_0 is μ. By (22), (24),

$$(26) \qquad \bar{\gamma}t = E^{\bar{u}}\left[\int_0^t I_G(\xi_s)k(\xi_s,\bar{u}_s)ds + \int_0^t I_{\partial G}(\xi_s)k_0(\xi_s,\bar{u}_s)ds\right].$$

Remarks. The limit process $\xi(\cdot)$ is stationary, as is the drift $\bar{f}(\cdot)$, but we have not been able to show that there is a Markov (reflecting diffusion) process with the same distributions. There probably is such a Markov process, as there probably is a stationary pure Markov control $\tilde{u}(\cdot)$ such that $\tilde{u}(\xi_t) = \bar{u}(\omega,t)$ w.p.1. In any case, our method gives much information on the optimal process $\xi(\cdot)$; e.g., the multivariate distribituions of $\xi^h(\cdot)$ converge weakly to those of $\xi(\cdot)$, as do the distributions of any bounded measurable functional $F(\xi^h(\cdot))$, if $F(x(\cdot))$ is continuous w.p.1 with the respect to the measure induced by $\xi(\cdot)$. Indeed, one of the great advantages of the weak convergence method is that it yields such information, in addition to approximations to $\bar{\gamma}$. Also, $\bar{\gamma}$ = average cost per unit time for $\xi(\cdot)$, and is the limit of the average costs per unit time for the sequence of approximations.

7. Optimality of the Limit $\xi(\cdot)$. Being a limit of optimal approximating processes, $\xi(\cdot)$ is a good candidate for optimality for the original optimization problem (with the reflected diffusion model). Certain optimality properties are easy to show.

Theorem 4. *Assume A1 and A2. Let $v(\cdot)$ denote a continuous stationary pure Markov control, such that the corresponding reflecting diffusion $\xi^v(\cdot)$ is unique (in the weak sense) and has a unique invariant measure μ^v. Then $\bar{\gamma} \le \gamma^v$ (where we let the initial measure be μ^v).*

Proof. Let $\hat{\xi}_n^h$ and $\hat{\xi}^h(\cdot)$ denote the discretized and interpolated processes, resp., corresponding to the fixed control $v(\cdot)$. Then the cost $\gamma^{v,h}$ for the interpolated process is $\geq \bar{\gamma}^h$ by optimality of u^h. Let $\mu^{v,h}$ denote any invariant measure for $\hat{\xi}^h(\cdot)$. Then $\{\hat{\xi}^h(\cdot)\}$ and the invariant measures $\{\mu^{v,h}\}$ converge weakly to $\xi^v(\cdot)$ and μ^v, resp., as $h \to 0$ by arguments similar to those in Theorems 1 to 2. The theorem follows from this and (24). Q.E.D.

Since we have not been able so far to prove that $\bar{u}(\cdot)$ is stationary pure Markov, it would be nice to prove that $\bar{u}(\cdot)$ is optimal with respect to a broader class of controls than those in Theorem 4. The class can be broadened, but at the expense of considerable terminology and detail. We refer the reader to [2], where broader classes of comparison controls are dealt with for a number of other types of optimization problems.

REFERENCES

[1] Kushner, H.J. (1978) Optimality conditions for the average cost per unit time problem with a diffusion model, SIAM J. Control and Optimization, 16, pp. 330-346.

[2] Kushner, H.J. (1977) Probability Methods for Approximations in Stochastic Control and for Elliptic Equations, Academic Press, New York.

[3] Kushner, H.J. (1976) Probabilistic methods for finite difference approximations to degenerate elliptic and parabolic equations with Neumann and Dirichlet boundary conditions, J. Math. Anal. and Applic., 53, pp. 643-688.

[4] Kushner, H.J. and DiMasi, G (1978) Approximations for functionals and optimal control problems on jump diffusion processes, J. Math. Anal. and Applic., 63, pp. 772-800.

[5] Strook, D.W. and Varadhan, S.R.S. (1971) Diffusion processes with boundary conditions, Comm. Pure Appl. Math., 24, pp. 147-225.

[6] Chung, K.L. (1960) Markov Chains with Stationary Transition Probabilities, Springer-Verlag, Berlin.

[7] Ross, S.M. (1968) Arbitrary state Markovian decision processes, Ann. Math. Stat. 39, pp. 2118-2122.

[8] Schweitzer, P.J. and Federgruen, A. (1977) The asymptotic behavior of undiscounted value iteration in Markov decision problems, Math. of Oper. Res. 2, pp. 360-381.

[9] Billingsley, B. (1968) Convergence of Probability Measures, Wiley, New York.

[10] Lindvall, T. (1973) Weak convergence of probability measures and random functions in the function space D[0,∞), J. Appl. Prob., 10, pp. 109-121.

[11] Skorokhod, A.V. (1956) Limit theorems for stochastic processes, Th. of Prob. and its Applic., 1, pp. 261-290 (Engl. Transl.).

Obtaining Approximate Solutions of Random Differential Equations by Means of the Method of Moments

Melvin D. Lax

Department of Mathematics
California State University at Long Beach, Long Beach, California 90840

1. Introduction
2. The Method of Moments
3. Random Initial Value Problems
4. Random Boundary Value Problems
5. Implementation
References

1. INTRODUCTION

In previous papers by Lax and Boyce [1] and Lax [2,3] the method of moments has been shown to be an effective technique for solving* linear random ordinary differential equations (RDE's) and linear random integral equations approximately. The purpose of this paper is to prove the validity of the method of moments for a wider class of RDE's than dealt with before, and also to provide an understandable introduction to the theory and application of this technique.

In addressing the first objective the results proved in [1] and [2] for random initial value problems (RIVP's) and random boundary value problems (RBVP's) having discretizable stochastic processes as coefficients are extended to RIVP's and RBVP's whose coefficients belong to a larger set of stochastic processes defined herein as "nice". In addition, a new theorem concerning the error involved in using truncated random Taylor series in solving RBVP's is given.

To help achieve the second objective, we restrict the discussion to 2nd order RDE's of the form**

$$Y'' + Q(t,\omega)Y = F(t,\omega).$$

This simplifies the exposition and focuses attention upon those RDE's most important in applications. Also a special effort has been made to include examples and computing information which would foster a feel for the implementation of the technique and for the type of results it can achieve.

2. THE METHOD OF MOMENTS

The method of moments (see Vorobyev [4]) is a modification of Galerkin's method which enables the known operator and functions in an operator equation to be used in constructing subspaces in which an approximate solution to the equation may be found. Its advantage over Galerkin's method is that usually a more accurate approximation is produced for subspaces of the same size. Indeed when viewed as an iterative method, the method of moments' primary asset is its speed of convergence.

---

*By "solving an RDE" we shall mean "find the mean and autocorrelation of the random solution."

**The reader should note that any linear 2nd order RDE can be transformed so that it appears in this form.

Consider the equation

(1) $$x = Ax + f$$

where A is a known operator and f is a known function in a Hilbert space H whose inner product is $(\cdot,\cdot)$. Choose $z_0 = f$. Define

(2) $$z_i = Az_{i-1} \quad \text{for} \quad i = 1, \ldots, n.$$

(If the z_i's are linearly dependent, problem (1) is easily solved, so in the following the more difficult and more common case where the z_i's are linearly independent is discussed.)

On the subspace H_n whose basis is $\{z_0, \ldots, z_{n-1}\}$, an approximating operator A_n may be defined so that

(3) $$A_n z_i = z_{i+1}, \quad i = 0, \ldots, n-2$$
$$A_n z_{n-1} = E_n z_n$$

where E_n is the projection operator from H onto H_n. Since $E_n z_n \in H_n$, there exist constants $c_0, c_1, \ldots, c_{n-1}$ such that

(4) $$E_n z_n = -c_0 z_0 - c_1 z_1 - \ldots - c_{n-1} z_{n-1}.$$

By taking the inner product of (4) with $z_0, \ldots, z_{n-1}$ successively and noting that $(E_n z_n, z_k) = (z_n, z_k)$ for $k = 0, \ldots, n-1$, the following system of equations is obtained:

(5) $$\begin{aligned} (z_0,z_0)c_0 + (z_0,z_1)c_1 + \ldots + (z_0,z_{n-1})c_{n-1} + (z_0,z_n) &= 0 \\ (z_1,z_0)c_0 + (z_1,z_1)c_1 + \ldots + (z_1,z_{n-1})c_{n-1} + (z_1,z_n) &= 0 \\ &\vdots \\ (z_{n-1},z_0)c_0 + (z_{n-1},z_1)c_1 + \ldots + (z_{n-1},z_{n-1})c_{n-1} + (z_{n-1},z_n) &= 0. \end{aligned}$$

Now the approximating equation

(6) $$x_n = A_n x_n + f$$

may be solved. Let $k_0, \ldots, k_{n-1}$ be constants such that

$$x_n = k_0 z_0 + k_1 z_1 + \ldots + k_{n-1} z_{n-1}.$$

Then (6) may be written as

$$z_0 + \sum_{i=0}^{n-1} (A_n - 1)k_i z_i = (1-k_0)z_0 + \sum_{i=1}^{n-1} (k_{i-1}-k_i)z_i - k_{n-1} \sum_{i=0}^{n-1} c_i z_i$$

$$= (1-k_0-k_{n-1}c_0)z_0 + \sum_{i=1}^{n-1} (k_{i-1}-k_i-k_{n-1}c_i)z_i$$

(7) $$= 0.$$

Since the z_i's are linearly independent, their coefficients must all be zero in (7). Hence

$$k_{n-1} = \frac{1-k_0}{c_0}$$

$$k_{i-1} = k_i + \left(\frac{1-k_0}{c_0}\right) c_i \,, \qquad i = 1, \ldots, n-1$$

which implies that

$$(8) \qquad k_{i-1} = \frac{1-k_0}{c_0}\left(1 + \sum_{j=i}^{n-1} c_j\right), \qquad i = 1, \ldots, n-1.$$

Solving (8) for $i = 1$ yields

$$(9) \qquad k_0 = 1 - \frac{c_0}{1+\sum_{j=0}^{n-1} c_j}\,.$$

The other k_i's can be easily obtained in explicit form, but for computer implementation the following recursion relation is more convenient:

$$(10) \qquad k_i = k_{i-1} - \frac{c_i}{1+\sum_{j=0}^{n-1} c_j}\,, \quad i = 1, \ldots, n-1.$$

Having solved (6) it remains to establish the validity of its solution as an approximation of the solution of (1). A well-known result (see [4], p. 28) may be stated as follows:

Theorem 1: Let A be a completely continuous operator. Let A_n be defined by equation (3). Let 1 be a regular value of problem (1). Then for n sufficiently large problem (6) has a solution x_n, and the sequence $\{x_n\}$ converges to $\dot{x}$, the solution of problem (1), in the sense that

$$\lim_{n\to\infty} \|x - x_n\| = 0,$$

where $\|\cdot\|$ is the induced norm of H.

---

3. RANDOM INITIAL VALUE PROBLEMS

Consider the m initial value problems

$$y'' + q_i\,(t)\,y = f_i\,(t),\ y(0) = y'(0) = 0$$

where q_i, f_i are continuous for $i = 1, \ldots, m$. If a probability of $a_i > 0$ is given to the ith problem, where $\sum_{i=1}^{m} a_i = 1$, and the stochastic processes $Q(t,\omega)$ and $F(t,\omega)$ are defined to have sample functions $q_i(t)$ and $f_i(t)$, respectively, then the above initial value problems may be written as the RIVP

(11) $Y'' + Q(t,\omega)Y = F(t,\omega), \quad Y(0,\omega) = Y'(0,\omega) = 0.$

The solution of problem (11) will be a stochastic process having m sample functions with respective probabilities $a_1, a_2, \ldots, a_m$.

To facilitate the use of the method of moments, RIVP (11) can be transformed into the random integral equation

$$Y''(t,\omega) = \int_0^t (s-t)\ Q(t,\omega)\ Y''(s,\omega)ds + F(t,\omega)$$

which in turn can be written as

$$Y'' = AY'' + F \tag{12}$$

where the operator A is defined by

$$AX = \int_0^t (s-t)\ Q(t,\omega)X\ ds\ .$$

The domain of A is the Hilbert space H whose elements are in the set

$$\{X(t,\omega) : X(t,\omega) \text{ is a stochastic process; } \omega = \omega_i \text{ with probability } a_i \text{ for } i = 1, \ldots, m; \; X(0,\omega) = X'(0,\omega) = 0; \; \int_0^1 E\{X^2(t,\omega)\}dt < \infty\}$$

and whose inner product is defined as

$$(X(t,\omega),Z(t,\omega)) = \int_0^1 E\{X(t,\omega)Z(t,\omega)\}dt.$$

The induced norm of this space will be denoted by

$$\|X(t,\omega)\|_H = \left(\int_0^1 E\{X^2(t,\omega)\}dt\right)^{1/2}$$

Lemma 1: The operator A is completely continuous in the Hilbert space H.

Proof: Let $\{X_k(t,\omega)\}$ be a sequence in H such that

$$\|X_k(t,\omega)\|_H^2 < C \tag{13}$$

for some constant $C > 0$. It suffices to produce a subsequence $\{X_{k_j}(t,\omega)\}$ and $V(t,\omega) \in H$ such that

$$\|AX_{k_j}(t,\omega) - V(t,\omega)\|_H^2 \to 0.$$

Equation (13) implies that

$$\sum_{i=1}^{m} a_i \int_0^1 X_k^2(t,\omega_i)dt < C$$

which in turn implies that for each i,

$$\int_0^1 X_k^2(t,\omega_i)dt < \frac{C}{\min\limits_{j=1,\ldots,m} a_j} \equiv C_0.$$

Since the operator $A^{(1)}$ defined by $A^{(1)}x(t) = \int_0^t (s-t)Q(t,\omega_1)x(s)ds$ is completely continuous in the (deterministic) Hilbert space $L_2([0,1])$, a subsequence $\{X_{k_{j_1}}(t,\omega)\}$ of $\{X_k(t,\omega)\}$ and a function $v_1(t)$ can be found for which

$$\int_0^1 \left[\int_0^t (s-t)Q(t,\omega_1)X_{k_{j_1}}(s,\omega_1)ds - v_1(t)\right]^2 dt \to 0.$$

Similarly a subsequence $\{X_{k_{j_2}}(t,\omega)\}$ of $\{X_{k_{j_1}}(t,\omega)\}$ and a function $v_2(t)$ can be found for which

$$\int_0^1 \left[\int_0^t (s-t)Q(t,\omega_2)X_{k_{j_2}}(s,\omega_2)ds - v_2(t)\right]^2 dt \to 0.$$

This procedure may be continued until at last a subsequence $\{X_{k_{j_m}}\}$ and functions $v_1, \ldots, v_m$ are obtained such that

$$\int_0^1 \left[\int_0^t (s-t)Q(t,\omega_i)X_{k_{j_m}}(s,\omega_i)ds - v_i(t)\right]^2 dt \to 0$$

for $i = 1, \ldots, m$. Hence if $V \in H$ is defined such that $V(t,\omega_i) = v_i(t)$, then $\|AX_{k_{j_m}}(t,\omega) - V(t,\omega_i)\|_H^2 \to 0$ as $m \to \infty$ which proves that A is completely continuous in H. ■

The above lemma and Theorem 1 taken together guarantee that the method of moments approximate solutions of problem (12) converge to the solution of problem (12). Applying the initial conditions of RIVP (11), we arrive at the following theorem.

Theorem 2: Let $\{\zeta_n\}$ be the sequence of method of moments solutions of problem (12). Let

$$Y_n(t,\omega) = \int_0^t \int_0^s \zeta_n(r,\omega)dr\, ds.$$

Let Y be the solution of RIVP (11). Then

$$\lim_{n\to\infty} \int_0^1 E\{[Y(t,\omega) - Y_n(t,\omega)]^2\}dt = 0.$$

Having verified that the method of moments may be used to approximate the solution of a RIVP whose coefficients are stochastic processes with a finite number of sample functions, we now seek to extend our result to a larger class of RIVP's. In the subsequent discussion we use the norm

$$\|P(t,\omega)\|_2 = E^{\frac{1}{2}}\{[P(t,\omega)]^2\}.$$

Definition 1: A stochastic process $P(t,\omega)$, $0 \le t \le 1$, is nice if

i) P has continuous sample functions.

ii) There exists a constant M such that $|P(t,\omega)| < M$.

iii) There exists a sequence $\{P_m(t,\omega)\}$ such that each P_m is a discrete stochastic process whose m sample functions are also sample functions of P, and $\lim_{m\to\infty} \|P(t,\omega) - P_m(t,\omega)\|_H = 0$, $0 \le t \le 1$.

Consider the RIVP

(14) $Y'' + Q(t,\omega)Y = F(t,\omega)$, $Y(0,\omega) = Y'(0,\omega) = 0$

where Q and F are nice.

Lemma 2: Let Y be the solution of RIVP (14). Then there exists a constant M_2 such that

$$|Y''(t,\omega)| < M_2.$$

Proof: Since Q and F are nice, there exists a constant M_1 such that

$$|Q(t,\omega)| < M_1$$

and

$$|F(t,\omega)| < M_1$$

Writing RIVP (14) as an integral equation, we have

$$Y''(t,\omega) = \int_0^t (s-t)Q(t,\omega)Y''(s,\omega)ds + F(t,\omega).$$

Therefore

$$|Y''(t,\omega)| \le \int_0^t M_1|Y''(s,\omega)|ds + M_1 .$$

By Gronwall's lemma,

$$|Y''(t,\omega)| \le M_1 e^{M_1 t} \le M_1 e^{M_1} = M_2 . \blacksquare$$

Let $M_3 = \max[M_2^2, M_2, 1]$. Since Q and F are nice, there exists m and also Q_m and F_m such that

$$\|Q(t,\omega) - Q_m(t,\omega)\|_H < \frac{\epsilon}{4M_3}, \quad 0 \le t \le 1$$

and

$$\|F(t,\omega) - F_m(t,\omega)\|_H < \frac{\epsilon}{4M_3}, \quad 0 \le t \le 1.$$

If Y is the solution of RIVP (14) and Y_m is the solution of

$$Y_m'' + Q_m(t,\omega)Y_m = F_m(t,\omega), \quad Y_m(0,\omega) = Y'_m(0,\omega) = 0$$

it follows that

$$Y''(t,\omega) - Y_m''(t,\omega)$$

$$= \int_0^t (s-t)Q(t,\omega)Y''(s,\omega)ds + F(t,\omega) - \int_0^t (s-t)Q_m(t,\omega)Y_m''(s,\omega)ds$$

$$- F_m(t,\omega)$$

$$= \int_0^t (s-t)[Q(t,\omega)Y''(s,\omega) - Q(t,\omega)Y_m''(s,\omega) + Q(t,\omega)Y_m''(s,\omega)$$

$$- Q_m(t,\omega)Y_m''(s,\omega)]ds + F(t,\omega) - F_m(t,\omega).$$

Now letting $\alpha = |Q-Q_m|$, $\beta = |F-F_m|$, and $\gamma = |Y''-Y_m''|$, we have

$$\gamma(t,\omega) \le \int_0^t M_1\, \gamma(s,\omega)ds + \int_0^t M_2\, \alpha(t,\omega)ds + \beta(t,\omega)$$

$$\le M_1 \int_0^t \gamma(s,\omega)ds + M_2\, \alpha(t,\omega) + \beta(t,\omega).$$

By Gronwall's lemma

$$\gamma(t,\omega) \le M_1 \int_0^t e^{M_1(t-s)} [M_2\alpha(s,\omega) + \beta(s,\omega)]ds$$

$$+ M_2\alpha(t,\omega) + \beta(t,\omega)$$

$$\le M_1 e^{M_1} M_2 \int_0^t \alpha(s,\omega)ds + M_1 e^{M_1} \int_0^t \beta(s,\omega)ds$$

$$+ M_2\alpha(t,\omega) + \beta(t,\omega).$$

By the triangle inequality and Hölder's inequality, it follows for $0 \le t \le 1$ that

$$\|\gamma\|_2 \le M_3 [\, \|\int_0^t \alpha(s,\omega)ds\|_2 + \|\int_0^t \beta(s,\omega)ds\|_2 + \|\alpha\|_2 + \|\beta\|_2 \,]$$

$$\le M_3 [\, (E \int_0^t \{\alpha^2(s,\omega)\}ds)^{\frac{1}{2}} + (E \int_0^t \{\beta^2(s,\omega)\}ds)^{\frac{1}{2}}$$

$$+ \|\alpha\|_2 + \|\beta\|_2 \,]$$

$$\leq M_3 \, [\|\alpha\|_H + \|\beta\|_H + \|\alpha\|_2 + \|\beta\|_2]$$

$$\leq M_3 \, [\frac{\epsilon}{2M_3} + \|\alpha\|_2 + \|\beta\|_2].$$

Hence the identity $\|X\|_H = \| \, \|X\|_2 \, \|_H$ and the triangle inequality imply that

$$\|\gamma\|_H \leq M_3 \, [\frac{\epsilon}{2M_3} + \frac{\epsilon}{4M_3} + \frac{\epsilon}{4M_3}] = \epsilon.$$

Combining the above result with Theorem 2 leads to the following key theorem.

Theorem 3: Let $Q(t,\omega)$ and $F(t,\omega)$ be nice stochastic processes. Let $\{Q_m(t,\omega)\}$ and $\{F_m(t,\omega)\}$ be as given in Definition 1. For each m, let $\{Y_{mn}(t,\omega)\}$ denote the sequence of solutions generated by the method of moments that converges to the solution $Y_m(t,\omega)$ of

$$Y'' + Q_m(t,\omega)Y = F_m(t,\omega), \quad Y(0,\omega) = Y'(0,\omega) = 0.$$

Then if $Y(t,\omega)$ is the solution of

$$Y'' + Q(t,\omega)Y = F(t,\omega), \quad Y(0,\omega) = Y'(0,\omega) = 0,$$

it follows that

$$\lim_{m\to\infty} \lim_{n\to\infty} \int_0^1 E\{[Y(t,\omega) - Y_{mn}(t,\omega)]^2\}dt = 0.$$

---

4. RANDOM BOUNDARY VALUE PROBLEMS

The results just developed for RIVP's can also be established for RBVP's of the form

$$(15) \qquad Y'' + Q(t,\omega)Y = F(t,\omega)$$
$$\alpha Y'(0,\omega) + \beta Y(0,\omega) = 0,$$
$$\gamma Y'(1,\omega) + \delta Y(1,\omega) = 0,$$

where $\delta\gamma \geq 0$, $\alpha\beta \leq 0$, $\beta\delta \neq 0$.

Let Q,F have a finite number of sample functions. Equation (15) can be written as

$$(16) \qquad A_0 Y = A_1 Y + F$$

where $A_0 Y = Y''$ and $A_1 Y = -QY$.

Since $A_0^{-1} X(t,\omega) = \int_0^1 G_0(t,s,\omega)X(s,\omega)ds$, where $G_0(t,s,\omega)$ is the Green's function of RBVP (15) with $Q = F = 0$, we may apply A_0^{-1} to (16) to get

$$(17) \qquad Y = AY + F_0$$

where $$AY = \int_0^1 G_0(t,s,\omega)Q(s,\omega)Y(s,\omega)ds$$

and $$F_0 = \int_0^1 G_0(t,s,\omega)F(s,\omega)ds.$$

With similar reasoning to that of the proof of Lemma 1, it can be shown that A is a completely continuous operator on the Hilbert space L whose elements are in the set $\{X(t,\omega):X(t,\omega)$ is a stochastic process; $\omega = \omega_i$ with probability a_i for $i = 1,\ldots,m$;

$$\alpha X'(0,\omega) + \beta X(0,\omega) = 0; \quad \gamma X'(1,\omega) + \delta X(1,\omega) = 0;$$

$$\int_0^1 E\{[X^{(k)}(t,\omega)]^2\}dt < \infty \quad \text{for} \quad k = 0,1,2\}$$

and whose inner product is defined as

$$(X(t,\omega),Z(t,\omega)) = -\int_0^1 E\{X''(t,\omega)Z(t,\omega)\}dt.$$

From this fact and Theorem 1 the following result is obtained.

Theorem 4: Let Y be the solution of RBVP (15) where Q,F have a finite number of sample functions. Let $\{Y_n\}$ be the sequence of method of moments solutions. Then

$$\lim_{n\to\infty} \int_0^1 E\{[Y(t,\omega) - Y_n(t,\omega)]^2\}dt = 0.$$

Now consider RBVP (15) where Q and F are nice stochastic processes. It can be written in the form

$$Y(t,\omega) = \int_0^1 G(t,s,\omega)F(s,\omega)ds \tag{18}$$

where $G(t,s,\omega)$ is the Green's function associated with RBVP (15). It is assumed that $\lambda = 0$ is not an eigenvalue of

$$\begin{aligned} Y'' + Q(t,\omega)Y &= \lambda Y \\ \alpha Y'(0,\omega) + \beta Y(0,\omega) &= 0 \\ \gamma Y'(1,\omega) + \delta Y(1,\omega) &= 0 \end{aligned} \tag{19}$$

and 0 is not the limit of any sequence of sample eigenvalues of the random eigenvalue problem (19). Then $G(t,s,\omega)$ may be expressed as

$$G(t,s,\omega) = \begin{cases} \dfrac{X(s,\omega)Z(t,\omega)}{W(X(t,\omega),Z(t,\omega))}, & 0 \le s \le t \\ \dfrac{X(t,\omega)Z(s,\omega)}{W(X(t,\omega),Z(t,\omega))}, & t \le s \le 1 \end{cases} \tag{20}$$

where X is the solution of the RIVP

$$(21) \qquad X'' + Q(t,\omega)X = 0, \quad X(0,\omega) = \alpha, \; X'(0,\omega) = -\beta$$

and $\quad Z$ is the solution of the RIVP

$$(22) \qquad Z'' + Q(t,\omega)Z = 0, \quad Z(1,\omega) = \gamma, \; Z'(1,\omega) = -\delta$$

and $\quad W$ is the Wronskian of X and Z

$$(23) \qquad W(X(t,\omega),Z(t,\omega)) = X(t,\omega)Z'(t,\omega) - X'(t,\omega)Z(t,\omega).$$

By Abel's identity W is constant with respect to t, so evaluating (23) at $t = 0$ yields

$$(24) \qquad W = \alpha Z'(0,\omega) + \beta Z(0,\omega).$$

Hence by the assumption of the previous paragraph

$$(25) \qquad \operatorname{ess}_\omega \inf W \geq M_1 > 0, \qquad 0 \leq t \leq 1,$$
$$\text{or} \qquad \operatorname{ess}_\omega \sup W \leq -M_1 < 0, \qquad 0 \leq t \leq 1,$$

where M_1 is a constant. It is also true that there is a constant M_2 such that $|X(t,\omega)|, |Z(t,\omega)|$, and $|W(X(t,\omega),Z(t,\omega))|$ are less than M_2.

Let $\epsilon > 0$ and $M_3 = \dfrac{M_1{}^2}{2M_2{}^3(\alpha+\beta+3)}$.

Then by Definition 1 and reasoning similar to the proof of Theorem 3, there exists m and also $Q_m(t,\omega)$ and $F_m(t,\omega)$ such that

$$\|Q(t,\omega) - Q_m(t,\omega)\|_H < \epsilon\, M_3$$

$$\|F(t,\omega) - F_m(t,\omega)\|_H < \epsilon\, M_3$$

$$\|X(t,\omega) - X_m(t,\omega)\|_H < \epsilon\, M_3$$

$$\|Z(t,\omega) - Z_m(t,\omega)\|_H < \epsilon\, M_3$$

$$\|Z'(t,\omega) - Z_m'(t,\omega)\|_H < \epsilon\, M_3$$

where X_m and Z_m are the solutions of RIVP (20) and (21), respectively, with Q_m replacing Q. The preceding analysis which produced equations (18), (20), (24), and (25) can be duplicated to yield similar equations involving F_m, G_m, W_m, X_m, Y_m, and Z_m. These equations imply that

$$|Y(t,\omega) - Y_m(t,\omega)| = \left|\int_0^1 [G(t,s,\omega)F(s,\omega) - G_m(t,s,\omega)F_m(s,\omega)]ds\right|$$

$$= \left|\int_0^t \left[\frac{X(s,\omega)Z(t,\omega)F(s,\omega)}{W} - \frac{X_m(s,\omega)Z_m(t,\omega)F_m(s,\omega)}{W_m}\right] ds\right.$$

$$+ \int_t^1 \left[\frac{X(t,\omega)Z(s,\omega)F(s,\omega)}{W} - \frac{X_m(s,\omega)Z_m(t,\omega)F_m(s,\omega)}{W_m} \right] ds \Bigg|$$

$$\leq \int_0^t \left| \frac{W_m XZF - WXZF}{WW_m} \right| ds + \int_0^t \left| \frac{WXZF - WX_m ZF}{WW_m} \right| ds$$

$$+ \int_0^t \left| \frac{WX_m ZF - WX_m Z_m F}{WW_m} \right| ds + \int_0^t \left| \frac{WX_m Z_m F - WX_m Z_m F_m}{WW_m} \right| ds$$

$$+ \int_t^1 \left| \frac{W_m XZF - WXZF}{WW_m} \right| ds + \int_t^1 \left| \frac{WXZF - WX_m ZF}{WW_m} \right| ds$$

$$+ \int_t^1 \left| \frac{WX_m ZF - WX_m Z_m F}{WW_m} \right| ds + \int_t^1 \left| \frac{WX_m Z_m F - WX_m Z_m F_m}{WW_m} \right| ds.$$

Reasoning as in the proof of Theorem 3, this inequality together with the triangle inequality and Hölder's inequality yields

$$\|Y(t,\omega) - Y_m(t,\omega)\|_H < \frac{2M_2^{\,3}\,\epsilon\, M_3(\alpha+\beta+3)}{M_1^{\,2}} = \epsilon.$$

The above result combined with Theorem 4 provide a key theorem for RBVP's which is analogous to Theorem 3 for RIVP's.

Theorem 5: Let $Q(t,\omega)$ and $F(t,\omega)$ be nice stochastic processes. Let $\{Q_m(t,\omega)\}$ and $\{F_m(t,\omega)\}$ be as given in Definition 1. For each m, let $\{Y_{mn}(t,\omega)\}$ denote the sequence of solutions generated by the method of moments that converges to the solution $Y_m(t,\omega)$ of

$$Y'' + Q_m(t,\omega)Y = F_m(t,\omega)$$
$$\alpha Y'(0,\omega) + \beta Y(0,\omega) = 0$$
$$\gamma Y'(0,\omega) + \delta Y(0,\omega) = 0.$$

Then if $Y(t,\omega)$ is the solution of

$$Y'' + Q(t,\omega)Y = F(t,\omega)$$
$$\alpha Y'(0,\omega) + \beta Y(0,\omega) = 0$$
$$\gamma Y'(0,\omega) + \delta Y(0,\omega) = 0,$$

and 0 is not an eigenvalue or the limit of a sequence of sample eigenvalues of the associated eigenvalue problem, it follows that

$$\lim_{m\to\infty} \lim_{n\to\infty} \int_0^1 E\{[Y(t,\omega) - Y_{mn}(t,\omega)]^2\}dt = 0.$$

5. IMPLEMENTATION

The preceding theoretical development relies upon a two step approach to solving RDE's. The solution of the original RDE is approximated by the solution of a RDE whose coefficients are stochastic processes with a finite number of sample functions. This intermediate approximating solution is then itself approximated via the method of moments. However upon actually applying this procedure to a RDE, one can well imagine the difficulty involved in producing the intermediate problem. Fortunately this can be avoided in practice because for sufficiently large m, the stochastic processes with m sample functions used in the intermediate problem can be selected so that their low order statistical moments are the same to computer accuracy as the low order statistical moments of the known stochastic processes of the original RDE.

Although one may diminish computational complexities by using the known stochastic processes in the original RDE, it may be necessary to further simplify the calculations. To this end truncated random Taylor series or truncated random Fourier series may be employed. The following theorem, which was proved in [1], justifies the use of a truncated random Taylor series in the RIVP and indicates an error bound on the resulting approximate solution.

Theorem 6: Let $Q(t,\omega)$ and $F(t,\omega)$ be stationary stochastic processes with respective autocorrelations $R_Q(\tau)$ and $R_F(\tau)$. Let $Q_n(t,\omega)$ and $F_n(t,\omega)$ be the n term truncated random Taylor series about zero approximating $Q(t,\omega)$ and $F(t,\omega)$, respectively. Let $|Q(t,\omega)| < M_2$ and $|F(t,\omega)| < M_3$ for $|t| < 1$. Let $Y(t,\omega)$ be the solution of

$$Y'' + Q(t,\omega)Y = F(t,\omega), \quad Y(0,\omega) = Y'(0,\omega) = 0$$

and let $Y_{nk}(t,\omega)$ be the solution of

$$Y'' + Q_n(t,\omega)Y = F_k(t,\omega), \quad Y(0,\omega) = Y'(0,\omega) = 0.$$

Then

$$\|Y(t,\omega) - Y_{nk}(t,\omega)\|_2$$

$$\le \left(\frac{M_5}{n!}|R_Q^{(2n)}(0)|^{\frac{1}{2}} + \frac{1}{k!}|R_F^{(2k)}(0)|^{\frac{1}{2}}\right)\left(\frac{e^{M_4|t|} - 1}{M_4}\right)$$

where $M_4 = \max(1, M_2)$ and $M_5 = \frac{M_3}{2M_2^2} e^{M_2}$.

A similar theorem can be proved for RBVP (15) under the assumption that the Wronskian W which appears in the proof of Theorem 5 is bounded sufficiently far away from 0. The proof involves a combination of the techniques used to prove Theorems 5 and 6. A statement of it for the boundary conditions $Y(0) = Y(1) = 0$ is given below.

Theorem 7: Let Q, F, R_Q, R_F, M_2, M_3, and M_4 be as defined in Theorem 6. Let M_1 be as defined in the proof of Theorem 5. Let

$M_5 = \frac{e^{M_2}}{2M_2}$ and $M_6 = \frac{(M_5+1)}{n!M_4} |R_Q^{(2n)}(0)|^{\frac{1}{2}}$.

Suppose that $M_1 > 2M_6(e^{M_4} - 1)$. Let $Y(t,\omega)$ be the solution of

$$Y'' + Q(t,\omega)Y = F(t,\omega), \quad Y(0,\omega) = Y(1,\omega) = 0$$

and let $Y_{nk}(t,\omega)$ be the solution of

$$Y'' + Q_n(t,\omega)Y = F_k(t,\omega), \quad Y(0,\omega) = Y(1,\omega) = 0.$$

Then

$$\begin{aligned} &\| Y(t,\omega) - Y_{nk}(t,\omega) \|_2 \\ &\le \frac{2M_5}{M_1^2} \Bigg[M_5 M_3 M_6 \left(e^{M_4} + e^{M_4 t} + e^{M_4(1-t)} - 3 \right) \\ &+ M_3 M_6^2 \left(e^{M_4 t} - 1 \right)\left(e^{M_4(1-t)} - 1 \right) \\ &+ \frac{1}{k!} \left(M_5 + M_6\left(e^{M_4 t} - 1 \right) \right) \left(M_5 + M_6 (e^{M_4(1-t)}) - 1 \right) \\ &\cdot \left| R_F^{(2k)}(0) \right|^{\frac{1}{2}} \Bigg]. \end{aligned}$$

A more concrete understanding of our technique for solving RDE's may be obtained by considering some simple examples.

Example 1.a.: $Y'' + V_1(\omega)Y = V_2(\omega) + V_3(\omega)t$, $Y(0,\omega) = Y'(0,\omega) = 0$ where V_1 and V_2 are independent, V_1 and V_3 are independent, V_1 is uniformly distributed on the interval $[2,4]$, $E\{V_2\} = 1$, $E\{V_2^2\} = 12$, $E\{V_3\} = 2$, $E\{V_3^2\} = 8$, and $E\{V_2 V_3\} = 4$.

This RIVP may be transformed to the random integral equation

$$Y''(t,\omega) = \int_0^t (s-t)V_1(\omega)Y''(s,\omega)ds + V_2(\omega) + V_3(\omega)t$$

which in turn may be written as the operator equation

$$Y''(t,\omega) = AY''(t,\omega) + V_2(\omega) + V_3(\omega)t$$

where $$AX(t,\omega) = \int_0^t (s-t)V_1(\omega)X(s,\omega)ds.$$

Setting $Z_0 = V_2 + V_3 t$ and using equation (2) we get

$$Z_1 = AZ_0 = \int_0^t (s-t)V_1(V_2+V_3 s)ds = -V_1\left(\frac{V_2t^2}{2} + \frac{V_3t^3}{6}\right)$$

$$\vdots$$

$$Z_n = AZ_{n-1} = \quad \cdots\cdots \quad = (-1)^n V_1^{\,n}\left(\frac{V_2t^{2n}}{(2n)!} + \frac{V_3t^{2n+1}}{(2n+1)!}\right).$$

The coefficients in system of equations (5) are

$$(Z_i,Z_j) = \int_0^1 E\left\{(-1)^{i+j}V_1^{\,i+j}\left(\frac{V_2t^{2i}}{(2i)!} + \frac{V_3t^{2i+1}}{(2i+1)!}\right)\left(\frac{V_2t^{2j}}{(2j)!} + \frac{V_3t^{2j+1}}{(2j+1)!}\right)\right\}dt$$

$$= (-1)^{i+j}E\{V_1^{\,i+j}\}\left[\frac{E\{V_2^{\,2}\}}{(2i)!(2j)!(2i+2j+1)} + \frac{E\{V_2V_3\}}{(2i+1)!(2j+1)!} + \frac{E\{V_3^{\,2}\}}{(2i+1)!(2j+1)!(2i+2j+3)}\right].$$

Using the first N basis functions, $Z_0,\ldots,Z_{N-1}$, we may solve system (5) and then plug into equations (9) and (10) to obtain $k_0,\ldots,k_{N-1}$. Thus the approximate solution of the RIVP is

$$\hat{Y}(t,\omega) = \int_0^t\int_0^s (k_0Z_0(r) + k_1Z_1(r) + \ldots + k_{N-1}Z_{N-1}(r))drds;$$

this may be used to find the desired approximations of the mean and autocorrelation.

This example was chosen, in part, because the actual solution may be readily found. It is

$$Y(t,\omega) = \frac{V_3t}{V_1} - \frac{V_3\sin(\sqrt{V_1}t)}{V_1^{\,3/2}\cos\sqrt{V_1}} + \frac{V_2}{V_1} - \frac{V_2}{V_1}\cos(\sqrt{V_1}t).$$

The actual and approximate means were calculated and compared at $t = .1, .2, \ldots, 1.0$ and found to be remarkably close for small N. Similarly excellent results were obtained for the autocorrelations at $(t_1,t_2) = (.1i,.1j)$ where $i = 1,\ldots,10$ and $j = 1,\ldots,10$. (See Table 1 and Table 2 below.) ●

Now let us examine an example where a RBVP is dealt with.

Example 1b: $Y'' + V_1(\omega)\, Y = V_2(\omega)\sin\pi t, \quad Y(0,\omega) = Y(1,\omega) = 0$

where V_1 and V_2 are as defined in Example 1a.

Although this RBVP may be expressed like operator equation (17), it is more convenient to work with if written in the form of operator equation (16). Thus we have

$$A_0 Z_n = A_1 Z_{n-1}, \quad Z_n(0,\omega) = Z_n(1,\omega) = 0,$$

that is,

$$Z_n'' = -V_1(\omega) Z_{n-1}, \quad Z_n(0,\omega) = Z_n(1,\omega) = 0.$$

The Z_0 is found by solving

$$A_0 Z_0 = V_2(\omega)\sin\pi t, \quad Z_0(0,\omega) = Z_0(1,\omega) = 0.$$

Hence

$$Z_0 = \frac{-V_2(\omega)}{\pi^2}\sin\pi t, \quad \text{and}$$

$$Z_n = -\frac{V_1{}^n(\omega)V_2(\omega)}{\pi^{2n+2}}\sin\pi t, \quad n \geq 0.$$

The coefficients of system (5) are found to be

$$(Z_i,Z_j) = -\int_0^1 E\{Z_i''Z_j\}dt = \frac{E\{V_1{}^{i+j}\}E\{V_2{}^2\}}{2\pi^{2i+2j+2}}.$$

Equations (5) (9) and (10) may be solved to get $k_0,\ldots,k_{N-1}$, so that an approximate solution is given by

$$\hat{Y}(t,\omega) = k_0 Z_0 + k_1 Z_1 + \ldots + k_{N-1} Z_{N-1}.$$

The actual solution of this example is

$$Y(t,\omega) = \frac{V_2\sin\pi t}{V_1+\pi^2}$$

The actual and approximate means were calculated at $t = .1i$ and the autocorrelations at $(t_1,t_2) + (.1i,.1j)$ where $i = 1,\ldots,9$ and $j = 1,\ldots,9$. Again excellent accuracy was obtained for small N. (See Table 1 below.) ●

In the preceding two examples it was possible to find a closed form expression for the general basis functions Z_n and their inner products (Z_i,Z_j). Usually this cannot be done, but instead a recursion formula for them can be found. In practice this causes no difficulty since these calculations are performed on a computer which, of course, finds a recursion formula to be just as good or better.*

Now let us examine data on the convergence of our scheme for the above example and also for the following RIVP's and RBVP's.

<u>Example 2</u>: $Y'' + V_1(\omega)Y = V_2(\omega)$

a. $Y(0,\omega) = Y'(0,\omega) = 0$

b. $Y(0,\omega) = Y(1,\omega) = 0$

where V_1 and V_2 are as defined in Example 1a.

<u>Example 3</u>: $Y'' + (V_1(\omega) + V_4(\omega)t)Y = V_2$

a. $Y(0,\omega) = Y'(0,\omega) = 0$

b. $Y(0,\omega) = Y(1,\omega) = 0$

where V_1 and V_2 are as defined in Example 1a and where V_4 is independent of V_1 and V_2 and uniformly distributed on the interval $[1,2]$.

Table 1 lists the number of digits to which the method of moments approximate solution agrees with the actual solution for Examples 1 and 2 and with a truncated perturbation expansion for Example 3 (since an exact closed form solution cannot be found.) The $\geq$ in the data for Example 3 indicates a closer agreement of successive method of moments solutions with each other than with the perturbation result (whose truncation has limited its accuracy).

*It should be noted that to perform the method of moments calculations it suffices to have just the 1st and 2nd order statistical moments of the stochastic processes appearing in the forcing function. For those stochastic processes appearing in the coefficient of Y, it suffices to know the $(2N-2)$th order statistical moments where N is the number of basis functions used. The modest amount of information needed about the given stochastic processes is an attractive feature of our approach.

TABLE 1
THE NUMBER OF ACCURATE DIGITS OBTAINED
BY THE METHOD OF MOMENTS

NUMBER OF BASIS FUNCTIONS USED	2	3	4	5	6
EXAMPLE					
1a, MEAN	1	3	5	7	9
AUTOCORRELATION	1	3	4	6	9
2a, MEAN	1	3	5	8	8
AUTOCORRELATION	1	3	4	7	8
3a, MEAN	1	3	4	≥5	
AUTOCORRELATION	1	2	≥4		
1b, MEAN	4	6	8	8	8
AUTOCORRELATION	3	5	8	8	8
2b, MEAN	2	3	5	6	7
AUTOCORRELATION	2	3	5	5	7
3b, MEAN	1	3	4	≥4	
AUTOCORRELATION	1	3	3	≥3	

Table 1 indicates that the methods of moments approximations for both RIVP's and RBVP's converge very rapidly to the desired mean and autocorrelation. In addition, about the same level of accuracy is maintained for all values of t considered. Table 2 provides evidence of this for Example 1a. The data listed in Table 2 are generated using six basis functions.

TABLE 2
NUMERICAL RESULTS FOR THE
MEAN IN EXAMPLE 1A

t	METHOD OF MOMENTS SOLUTION	ACTUAL SOLUTION	ERROR
0.1	0.00532034273596	0.00532034274444	-0.00000000000848
0.2	0.02245153888482	0.02245153891054	-0.00000000002572
0.3	0.05287619020251	0.05287619023749	-0.00000000003498
0.4	0.09767995739973	0.09767995742990	-0.00000000003016
0.5	0.15751994770660	0.15751994773006	-0.00000000002347
0.6	0.23260646867227	0.23260646870183	-0.00000000002956
0.7	0.32269871199831	0.32269871204366	-0.00000000004536
0.8	0.42711378570035	0.42711378575732	-0.00000000005698
0.9	0.54475001410175	0.54475001416751	-0.00000000006576
1.0	0.67412212115989	0.67412212123576	-0.00000000007588

The calculations in solving Examples 1, 2, 3 were done on an IBM 360/50 computer. The computer programs used were written in FORTRAN IV and compiled by a WATFIV compiler. As may be seen in Table 3, the amount of computer time required to solve the examples was small despite the fact that little effort was expended to make the computer programs efficient.* (It should be noted that the use of truncated random Taylor series to approximate known stochastic processes can be very efficacious in reducing programming difficulties and job times. In particular, this aids the computing of the basis functions, a task which poses the greatest programming obstacle. Also an important precaution is to utilize a good subroutine for calculating the solution of system of equations (5) as it may be rather ill-conditioned.)

*In Table 3 the unavailability of the missing data and the non-uniformity of the type of data presented are due to differences in the programs used to do the examples.

TABLE 3
COMPUTER TIME USED

EXAMPLE	COMPILE TIME (sec)	EXECUTION TIME (sec)	JOB TIME (sec)
1a, (JUST N=6)	4.27	22.82	30.04
2a, MEAN	3.95	2.69	9.50
AUTOCORRELATION	3.40	23.13	29.49
3a, MEAN	3.37	33.24	38.94
AUTOCORRELATION	--	--	300.00
1b, MEAN	2.74	6.63	12.60
AUTOCORRELATION	--	--	33.59
2b	4.75	24.22	33.00
3b, MEAN	5.40	127.93	138.40
AUTOCORRELATION	--	--	300.00

There is a major pitfall which must be avoided when using the method of moments to solve RBVP's. As may be seen in the proof of Theorem 5, it is necessary to avoid having an eigenvalue in the range of the stochastic process Q of RBVP (15). In practice it is also necessary to restrict the range of Q to be well away from such an eigenvalue in order to retain the rapid convergence found in the above examples. The following example demonstrates the strong effect that proximity to an eigenvalue has upon a method of moments solution.

Example 4: $Y'' + W(\omega)Y = V(\omega)$, $Y(0,\omega) = Y'(1,\omega) = 0$, where V and W are independent random variables, $E\{V\} = 1$, $E\{V^2\} = 12$, and W is uniformly distributed on the interval $[a,b]$.

Since $\lambda = \frac{\pi^2}{4} \approx 2.47$ is an eigenvalue of the deterministic eigenvalue problem

$$y'' + \lambda y = 0, \quad y(0) = y'(1) = 0,$$

we consider intervals $[a,b]$ close to 2.47. As may be seen in Table 4, there appears to be no convergence at all when close to the eigenvalue, and even when decently separated, the convergence is slower than normal.

TABLE 4

THE NUMBER OF ACCURATE DIGITS OBTAINED FOR THE MEAN IN EXAMPLE 4 BY THE METHOD OF MOMENTS

NUMBER OF BASIS FUNCTIONS USED	2	3	4	5	6
[a,b]					
[2,5,4.5]	0	0	0	0	0
[2.52,4.52]	0	0	1	0	1
[2.55,4.55]	0	0	1	0	1
[2.6,4.6]	0	0	1	1	1
[2.7,4.7]	0	0	1	1	1
[3,5]	0	1	1	1	2
[4,6]	0	2	1	2	3

Despite this rather pronounced sensitivity to eigenvalues, the method of moments approach remains a viable technique for solving RDE's -- a technique capable of exceptional accuracy when implemented on a computer. Similar success has been achieved for random integral equations (see [3]) and currently work is underway to apply the method of moments to random partial differential equations and random eigenvalue problems.

REFERENCES

1. Lax, M. D. and Boyce, W.E., "The Method of Moments for Linear Random Initial Value Problems," J. Math.Anal. Appl. 53 (1976), 111-132.

2. Lax, M.D., "The Method of Moments for Linear Random Boundary Value Problems," SIAM J. Appl. Math. 31 (1976), 62-83.

3. Lax, M.D., "Method of Moments Approximate Solutions of Random Linear Integral Equations," J. Math. Anal. Appl. 58 (1977), 46-55.

4. Vorobyev, Y.V., Method of Moments in Applied Mathematics, Gordon and Breach, New York, 1965.

Random Generalized Inverses and Approximate Solutions of Random Operator Equations

M. Z. Nashed and H. W. Engl*

Department of Mathematical Sciences
University of Delaware, Newark, Delaware 19711

Dedicated to Albert Bharucha-Reid on the occasion of his 50th birthday.

We see that the theory of probability is at bottom only common sense reduced to calculations; it makes us appreciate with exactitude what reasonable minds feel by a sort of instinct, often without being able to account for it . . . . It is remarkable that this science, which originated with the consideration of games of chance should have become the most important object of human knowledge . . . . The most important questions of life are, for the most part, really only problems of probability.

Pierre Simon, Marquis de Laplace

Nothing in Nature is random . . . . A thing appears random only through the incompleteness of our knowledge.

Benedictus de Spinoza, *Ethics* I

*On leave from Institut für Mathematik, Johannes-Kepler-Universität, A-4045 Linz, Austria.

I. Fundamental Concepts and Tools for Theory and Approximation of Random Operator Equations
1. Random Operators
2. Separable Random Operators
3. Techniques for Establishing Measurability of Solutions of Random Operator Equations
4. Measurability of Inverses of Random Operators
5. Measurability of Adjoints of Random Linear Operators
6. Measurability of Projectors

II. Generalized Inverses of Random Linear Operators
1. Generalized Inverses of Linear Operators in Banach and Hilbert Spaces
2. Series and Integral Representations of Generalized Inverses of Linear Operators
3. Tools for Establishing Measurability of Generalized Inverses of Linear Operators in Hilbert and Banach Spaces
4. Measurability of Generalized Inverses of Random Linear Operators Between Hilbert Spaces
5. Generalized Inverses of Random Bounded Linear Operators in Banach Space
6. Measurability of Outer Inverses

III. Iterative and Projection Methods for Least-Squares Solutions of Linear Operator Equations in Hilbert Space
1. Iterative Methods
2. Projection Methods

IV. Applications to Random Integral, Differential, and Operator Equations
1. The Interplay Between Integral Operators and Their Generalized Inverses in the Spaces $L_2[0,1]$ and $C[0,11]$
2. Random Resolvents and Pseudoresolvents; Explicit Representation of the Random Operator $(I\text{-}\lambda\ (\omega)\ K(\omega))^\dagger$
3. Random Generalized Green's Function for an nth Order Random Linear Differential Operator
4. Iterative Methods for Random Linear Boundary-Value Problems
5. Approximations to Least-Squares Solutions of Random Integral Equations of the First and Second Kind
6. Applications to Nonlinear Problems

V. References

INTRODUCTION

The theory of random equations has developed along two major lines, one initiated by Itô and the other by the Prague school of probabilists around Špaček and Hanš. This paper is entirely in the framework of the latter approach. A "random equation" in this context is an equation where certain components, such as coefficients, kernels, inhomogenuous terms, boundary conditions, etc. are random variables or stochastic processes. Perspectives on these two approaches (and variants thereof) and their relations can be found in the books by Bharucha-Reid [7], Soong [71], and Tsokos and Padgett [73].

The main questions concerning random operator equations are essentially the same as those for deterministic operator equations, viz., questions of _existence_, _uniqueness_, _characterization_, _construction_ and _approximation_ of solutions (more generally, of least-squares or best approximate solutions, or of weak solutions in some sense). For random operator equations these questions have soft and hard aspects. The _soft_ aspects concern wide-sense solutions of random operator equations, i.e., mappings from the underlying probability space into the domain space of the operator which satisfy almost surely the random operator equation (in an appropriate notion of "solution"). Wide-sense solutions (or approximations thereof) do not require any special consideration from the standpoint of probabilistic analysis; all aspects mentioned above follow immediately from the deterministic counterpart. In other words, the soft aspects involve those aspects that are strictly deterministic; the appearance of a random parameter notwithstanding.

In most applications of random operator equations, one is interested in statistical properties (e.g., probability distribution, mean and variance of a solution, etc.) and therefore it is necessary to seek solutions that are measurable functions of the random parameter; this is an important aspect where the study of random operator equations takes a new and nontrivial dimension and departs from the deterministic theory.

The main three "_hard_" aspects of random operator equations are

(i) measure-theoretic aspects: random operators, existence of measurable solutions, properties of the set of all measurable solutions as a subset of the set of all wide-sense solutions.

(ii) probabilistic and statistical aspects: probability distributions and properties of functionals of random solutions, such as moments.

(iii) "experience theory" and modeling: estimates for the difference between the mean-value of solutions of the random equation and the (deterministic) solution of the "averaged equation"; in other words: How well does a deterministic model approximate the stochastic problem?

The surge in activity in probabilistic functional analysis over the last two decades is mainly due to three factors or trends which continue to influence developments in this field. The first is the impact of abstract methods, in particular functional-analytic and measure-theoretic methods, on the theory and approximation of random operator equations. The second factor is the increasing demand for approximation methods for solving random operator equations. Recent advances in numerical analysis and approximation theory have been brought to bear on some of the problems in random operator equations. Functional analysis has been very powerful in developing and unifying the theory of numerical methods (for a historical perspective, see the review by M. Z. Nashed in Bull. Amer. Math. Soc. 82 (1976) pp. 825-834). The third factor comes from technological problems which give rise to stochastic mathematical models. This factor has given more impetus to the study of random equations and has put the various approaches to a critical test; as the horizon of applications of random equations expanded, a better understanding was gained of the limitations of the approaches that are used. In particular, the approach due to Špaček and Hanš might be too general in certain contexts.

In the past decade, the thrust of research in the area of probabilistic functional analysis in the direction initiated by the Prague school has taken place along four fronts:

(1) The study of measurability of adjoints, inverses, and generalized inverses for linear random operators; applications of these results to the development of random analogues of inverse and implicit function theorems: Hanš [28], Bharucha-Reid [7], Nashed and Salehi [62], Nashed [56] and others.

(2) Random analogues of classical fixed point theorems in nonlinear functional analysis: Andrus and Nishiura [2], Bharucha-Reid [8], Engl [15], [19], Itoh [34], Kannan and Salehi [39], Lee and Padgett [47], Nowak [65] and others.

(3) Constructive and approximation schemes for deterministic operators in the setting of random operators.

(4) Applications to random differential and integral equations; adaptation of the theory to specific models in applications. This is widely discussed in [7], [71], [73]; see also [52] for another interesting application.

Rather than describing the paper section by section, we refer to the table of contents. However, it seems to be appropriate to make some general remarks about the scope of the results obtained. As mentioned above, random integral equations of the type we study are discussed in [7]. In this paper, we develop several tools which prove to be useful in answering many of the questions raised there, both in the context of existence and of approximation of random solutions. Also, we extend many of the results in [7] by dropping some stringent technical conditions that were used in the early development of the subject. For example, we considerably extend the scope of inversion theorems for random operators and prove results about measurability of singular values of compact random operators. Also, the study in [7] is restricted to integral equations of the second kind. We give a parallel treatment of random operator equations with closed and nonclosed range; this includes as special cases integral equations of the second and first kind, respectively. The techniques needed and results obtained for these two types of equations are markedly different in the deterministic case (see [55]); these differences manifest themselves in the methods used to obtain random analogues.

The major part of this paper deals with random operator equations which are not necessarily uniquely solvable. A natural setting then is to consider these equations in the framework of generalized inverses, thereby studying least-squares and projectional solutions. This provides our motivation for investigating measurability of various kinds of generalized inverses of linear random operators. However, the need for random generalized inverses arises also in other contexts (e.g., in the study of infinite-dimensional stationary stochastic processes, the factorization problem for nonnegative definite operator-valued functions; see, e.g., Section U in the Annotated Bibliography [61]).

Approximation methods play two roles in the treatment of random operator equations. Clearly such methods are needed when explicit solutions are difficult to obtain. They also provide an effective approach to establishing measurability of solutions (see Section I.3). In this paper we provide a general framework which makes it possible to adapt to random equations approximation schemes that have proved effective for operator equations. This framework makes it possible in a routine way to impart to the field of random operator equations the powerful methods and results of numerical functional analysis. We illustrate this process by several specific iterative and projectional methods.

In Section IV we show how the abstract results about existence and approximation of solutions of random operator equations can be applied to a variety of problems for random differential and integral equations. We treat solvability in the framework of least-squares solutions. It should be pointed out that here as in Sections II and III most of the results are new also in the case of solvable or even uniquely solvable equations.

In a recent book review, Barry M. Mitchell remarked: "Somewhere between saying too little and saying too much lies good exposition. Much of the pitfalls are located to one side or the other of that rather narrow ridge where essential ideas are provided without a deluge of trivialities." It has been our objective to provide a framework within which random analogues of specific results can be obtained. We hope we have sufficiently delineated this framework without falling off this ridge.

The road from consideration of games of chance, where probability originated, to the contemporary developments in probabilistic functional analysis has been long and interesting. Along this road there are many landmarks reflecting major developments which opened new tracks. It would be fair to say that there was always a major track for applications of probability theory, and that applied problems have provided motivation for some of the important theoretical developments. Although some of the technical aspects of probability theory have acquired a very abstract level, it is true now as it was at the time of the famous marquis that "common sense reduced to calculations" is at the bottom of many developments in probability theory. In the realm of random operator equations, the subject of this paper, the increasing need for probabilistic methods in applied mathematics has led simultaneously to the consideration of random transformations on various algebraic and topological sturctures and to the development of approximation and numerical methods for random operator equations. Operator theory, measure theory, numerical functional analysis, and the theory of stochastic processes provide basic tools for research in the field of random operator equations. Several aspects of the interplay between these fields and random operator equations will be demonstrated and extended in this paper.

I. FUNDAMENTAL CONCEPTS AND TOOLS FOR THEORY AND APPROXIMATION OF RANDOM OPERATOR EQUATIONS

Unless stated otherwise, throughout this paper $(\Omega, \mathcal{A}, \mu)$ will be a complete σ-finite measure space and X,Y separable Banach spaces.

1. Random Operators

Definition 1.1: $C: \Omega \to \{A \subseteq X: A \neq \emptyset\}$ is called "measurable" if for all open $D \subseteq X$, $\{\omega \in \Omega: C(\omega) \cap D \neq \emptyset\} \in \mathcal{A}$. The "graph of C" is defined as

$$\text{Gr } C := \{(\omega,x) \in \Omega \times X: x \in C(\omega)\}.$$

If all $C(\omega)$ are singletons $\{x(\omega)\}$, we will identify $\{x(\omega)\}$ with $x(\omega)$, so that Def. 1.1 also defines measurability for $x: \Omega \to X$. Measurable functions from Ω into X will also be called (generalized) random variables, even if $\mu(\Omega) \neq 1$. This is a straightforward generalization of real-valued random variables.

Conceptually, a "random operator" is a family of operators from X into Y depending on $\omega \in \Omega$ such that it maps each $x \in X$ into a random variable (instead of a single element) in Y. Because we want to permit that the realizations of a random operator for any fixed value of ω, $T(\omega)$, are defined on different sets, say $C(\omega)$, we choose the graph of a set-valued map C (cf. Def. 1.1) as the domain of a random operator. By 2^X we denote $\{A \subseteq X: A \neq \emptyset$ and A closed$\}$. For $T(\omega,x)$ we also write $T(\omega)x$. For a random operator T we will denote by $D(T(\omega))$, $R(T(\omega))$, $N(T(\omega))$ the domain, range, and kernel (null space) of $T(\omega)$. Whenever we write "$T(\omega,x)$" we will assume that $x \in D(T(\omega))$.

Definition 1.2: Let $C: \Omega \to \{\emptyset \neq A \subseteq X\}$. $T: \text{Gr } C \to Y$ will be called "random operator from $\Omega \times X$ into Y" if for all $x \in X$ and open $D \subseteq Y$,

$$\{\omega \in \Omega: x \in C(\omega) \text{ and } T(\omega,x) \in D\} \in \mathcal{A}.$$

If in addition C is measurable, T is called "random operator with stochastic domain C". If $C(\omega) = X$ for all $\omega \in \Omega$, T is called "random operator on $\Omega \times X$". Such a T will be called "linear, bounded, continuous,..." if every $T(\omega,\cdot)$ is linear, bounded, continuous... .

Although phrased differently, this definition of a random operator from $\Omega \times X$ into Y coincides with Def. 1.2 of [62]. Note that if the domain $C(\omega)$ is dense in X for all $\omega \in \Omega$, then C is measurable, since then for every open set D ,

$\{\omega \in \Omega : C(\omega) \cap D \neq \emptyset\} = \Omega$.

The following result about the composition of random operators will be used frequently:

Lemma 1.3: Let in addition to X and Y also Z be a separable Banach space, $C: \Omega \to 2^X$ measurable such that there exists a countable $S \subseteq X$ with $\overline{S \cap C(\omega)} = C(\omega)$ for all $\omega \in \Omega$ ("separability assumption"), T a continuous random operator with stochastic domain C and values in Y, U a random operator from $\Omega \times Z$ into X such that for all $\omega \in \Omega$, $U(\omega)(D(U(\omega))) \subseteq C(\omega)$. Let $z: \Omega \to X$ be measurable such that for all $\omega \in \Omega$, $z(\omega) \in C(\omega)$. Then:

a) $\omega \to T(\omega, z(\omega))$ is measurable.

b) $T \circ U$ (mapping each (ω, z) into $T(\omega, U(\omega, z))$) is a random operator from $\Omega \times Z$ into Y.

Proof: As the proof of Lemma 10 in [16] with slight, but obvious modifications, cf. also [15]. □

For the "separability assumption" on C in Lemma 1.3, see, e.g., [20].

Although we do not assume separability of the underlying Banach spaces for the deterministic results we prove, we restrict our interest to separable spaces as soon as we talk about measurability for the following reason: Separable Banach spaces have the cardinality of the continuum. Random variables with values in spaces of higher cardinality, however, have the strange property that the sum of two of them need not be a random variable ([64]); this makes it doubtful if random variables with values in such spaces can serve as models for any reasonable real-world problem.

In all the results below "for all ω" in the assumptions can be replaced by "for μ-almost all ω" if we make the same change also in the conclusions.

Most results can be carried over to the case where (Ω, A, μ) is not necessarily complete. The conclusions, however, may hold then only for μ-almost all $\omega \in \Omega$ even if the assumptions are stated for <u>all</u> $\omega \in \Omega$. For the technique involved in the reduction of the general case to the case of a non-complete measure space see, e.g., [20].

2. Separable Random Operators

A special kind of stochastic processes which has been studied, e.g., in [24] is the class of separable processes. A similar notion of separability turns out to be useful also for random operators:

Definition 1.4: A random operator T from $\Omega \times X$ into Y is called "separable" if there exists a countable dense set $S \subseteq X$ and an $N \in \mathcal{A}$ with $\mu(N) = 0$ such that for every closed $K \subseteq Y$ and open $F \subseteq X$ we have

$$\{\omega \in \Omega: T(\omega,F \cap S) \subseteq K\} \backslash \{\omega \in \Omega: T(\omega,F) \subseteq K\} \subseteq N.$$

Any such S is called a "separant" of T.

Lemma 1.5: Let T be a random operator from $\Omega \times X$ into Y, S a countable dense subset of X. Then T is separable with separant S if and only if there exists an $N \in \mathcal{A}$ with $\mu(N) = 0$ such that for all $\omega \notin N$ and $x \in X$, there exists a sequence (x_i) of elements of S with $(x_i) \to x$ and $T(\omega, x_i) \to T(\omega,x)$.

Proof: $\Leftarrow$: Let $\omega \notin N$, $K \subseteq Y$ closed, $F \subseteq X$ open. We show that $T(\omega,F \cap S) \subseteq K$ implies $T(\omega,F) \subseteq K$. Let $x \in F$ be such that $T(\omega,x)$ is defined. By assumption there exists a sequence $(x_i) \in S$ (which may be assumed to be in $F \cap S$) such that $(x_i) \to x$ and $T(\omega,x_i) \to T(\omega,x)$. So if we assume $T(\omega,F \cap S) \subseteq K$, we have $T(\omega,x) \in K$ because of the closedness of K. As $x \in F$ was arbitrary, $T(\omega,F) \subseteq K$.

$\Rightarrow$: Let $\omega \notin N$, $x \in D(T(\omega))$. For $n \in \mathbb{N}$ let $F_n :=$ $\{z \in X: ||z-x|| < \frac{1}{n}\}$, $K_n := T(\omega,T_n \cap S)$. Since $\omega \notin N$ and $T(\omega,F_n \cap S) \subseteq K_n$, we have $T(\omega,F_n) \subseteq K_n$, especially $T(\omega,x) \in K_n$ for all $n \in \mathbb{N}$. By definition of K_n, there exists a sequence $(y_{in})_{i\in N}$ in $T(\omega,F_n \cap S)$ with $\lim\limits_{i\to\infty} y_{in} = T(\omega,x)$.

Let $\bar{x}_{in} \in F_n \cap S$ be such that $T(\omega,\bar{x}_{in}) = y_{in}$; for each $n \in \mathbb{N}$ we choose $i(n) \in \mathbb{N}$ with $i(n) \geq n$ and $||y_{i(n)n} - T(\omega,x)|| \leq \frac{1}{n}$ and define $x_n := \bar{x}_{i(n)n}$. Then $\lim\limits_{n\to\infty} x_n = x$ and $\lim\limits_{n\to\infty} T(\omega,x_n) = T(\omega,x)$. □

Corollary 1.6: Let T be a continuous random operator on $\Omega \times X$ into Y. Then T is separable.

Proof: Let S be a countable dense subset of X, $\omega \in \Omega$, $x \in X$. Then there is a sequence (x_i) in S converging to x. By continuity, $T(\omega,x_i) \to T(\omega,x)$. The assertion follows now from Lemma 1.5. □

A continuous random operator with stochastic domain $D(T(\omega))$ need not be separable, unless the stochastic domain fulfills the "separability assumption" of Lemma 1.3.

3. Techniques for Establishing Measurability of Solutions of Random Operator Equations

The earliest results on measurability of operators seem to be due to Hanš [28],[29], who considered some measurability problems for inverse and adjoint operator of a bounded linear random operator and gave a probabilistic analogue of the Banach contraction principle. The techniques used by Hanš may be regarded as ad hoc methods in the sense that one cannot extract from them general principles for establishing measurability for a wider class of problems. In recent years two basic general approaches have been used for establishing measurability of solutions of operator equations.

The idea of the first approach is based on representing the solution by a convergent approximation scheme (e.g., iterative and projection methods, series or integral representations). The key step is then to establish measurability of the approximants. Once this is accomplished, the measurability of the solution is an immediate consequence of the following lemma on limits:

Lemma 1.7. Let $\{x_n\}$ be a sequence of measurable functions from Ω into X converging (weakly or in the norm) to x ; then x is measurable.

Proof: follows from the results in Chapter 1 of [7] and the analogous result for $X = \mathbb{R}$. □

The advantage of this approach, which we will call the "limit theorem approach", is of course its constructive aspect.

If approximation schemes are not known, which is frequently the case when the solution is not unique, we resort to a "selection theorem approach". Before we describe this approach, we observe that in the case where the equation studied may have a non-unique solution, one cannot expect measurability of "all" solutions even in very simple cases as the following example shows. Let $(\Omega, A, \mu) := [0,1]$ with Lebesque measure, which is a complete probability space, $T: \Omega \times \mathbb{R} \to \mathbb{R}$ be defined by $T(\omega, x) := x^2 - 1$. This function can be viewed (artificially) as a random operator. Let $C \subset [0,1]$ be not Lebesque-measurable (C exists if we assume the validity of the axiom of choice). Then $x: \Omega \to \mathbb{R}$ defined by $x(\omega) = 1$ if $\omega \in C$ and $x(\omega) = -1$ if $\omega \in \Omega \setminus C$ is not measureable, but for each $\omega \in \Omega, T(\omega, x(\omega)) = 0$. So the "random" operator equation $T(\omega,x) = 0$ has a non-measurable solution. This pathology indicates that in the case of non-unique solutions the proper question to ask is not "Is every solution measurable?" but rather "Is there a measurable solution?" and possibly: "Are there measurable solutions dense in the set of all solutions?"

Selection theorems play a useful role in answering these questions. The key step in the selection theorem approach is to establish the measurability of the set-valued map (Def. 1.1) which takes each $\omega \in \Omega$ into the solution set of the realization of the equation corresponding to this ω. To this set-valued map one applies a selection theorem like Theorem 1.8 to conclude the existence of single-valued measurable maps which are solutions of the equation for each ω.

Theorem 1.8. Let $C: \Omega \to 2^X$. Then C is measurable if and only if there exists a countable set of measurable functions $z_1, z_2, z_3, \ldots : \Omega \to X$ such that for all $\omega \in \Omega$, $C(\omega) = \overline{\{z_1(\omega), z_2(\omega), z_3(\omega), \ldots\}}$.

For a proof see, e.g., Himmelberg [31].

The z_i's are called "measurable selectors". The existence of at least one measurable selector is the content of the classical Kuratowski-Ryll-Nardzewski selection theorem [43]. The countable dense set of measurable selectors as it appears in Theorem 1.8 is sometimes called a "Castaign representation of C." In this paper Theorem 1.8 essentially is the only selection theorem we use; however there are numerous selection theorems for more general situations (cf. Wagner [74], Ioffe [33]; for an interesting connection with "continuous selection theorems" we refer to Mägerl [41]). We remark that sometimes selection theorems have to be used to establish measurability of the approximants in the limit theorem approach, so that we actually use a combination of both techniques.

4. Measurability of Inverses of Random Operators

Theorem 1.9 (cf. Nashed and Salehi [62]). Let T be a separable random (not necessarily linear) operator from $\Omega \times X$ into Y. Then $R(T(\cdot))$ and $\overline{R(T(\cdot))}$ are measurable. Moreover, if $T(\omega,\cdot)$ is invertible for all ω and its inverse is continuous, then T^{-1} is also a random operator from $\Omega \times Y$ into X.

Proof. The measurability of $R(T(\cdot))$ follows from Lemma 1.5. The measurability of $\overline{R(T(\cdot))}$ follows then from [31]. Let the superscript c denote the complementation, $S(\cdot,r)$ and $\overline{S}(\cdot,r)$ denote the open and closed balls of radius r around $\cdot$ respectively.

To prove the measurability of T^{-1} it suffices to show that for an arbitrary $y \in Y$ and a closed ball $\overline{S}(x',r)$ the event $\{\omega: T^{-1}(\omega,y) \in \overline{S}(x',r)\} \in \mathcal{A}$. We note that

$$\{\omega: T^{-1}(\omega,y) \in \overline{S}(x',r)\} = \bigcup_{x\in\overline{S}(x',r)} \{\omega: x\in D(T(\omega)) \text{ and } T(\omega,x) = y\}.$$

We assert that

$$(1.1)\quad \bigcup_{x\in\overline{S}(x',r)} \{\omega: x\in D(T(\omega)) \text{ and } T(\omega,x) = y\} =$$

$$\bigcap_{n=1}^{\infty} \bigcup_{x\in S(x',r+\frac{1}{n})} \{\omega: x\in D(T(\omega)) \text{ and } T(\omega,x) \in S(y,\tfrac{1}{n})\}.$$

It suffices to show that the right-hand side is contained in the left-hand side. Let $\omega_o \in \bigcap_{n=1}^{\infty} \bigcup_{S(x',r+\frac{1}{n})} \{\omega: x\in D(T(\omega))$ and $T(\omega,x) \in S(y,\frac{1}{n})\}$. Then for each n, there exists $x_n \in S(x',r+\frac{1}{n})$ such that $T(\omega_o,x_n) \in S(y,\frac{1}{n})$. It follows that $\lim_{n\to\infty} T(\omega_o,x_n) = y$. Since $T^{-1}(\omega_o,\cdot)$ is continuous it follows that $\{x_n\}$ converges to $T^{-1}(\omega_o,y)$, which we denote by x. Hence $x \in D(T(\omega))$ and $T(\omega_o,x) = y$. Moreover $x \in \overline{S}(x',r)$. This implies that $\omega_o \in \bigcup_{\overline{S}(x',r)} \{\omega: x \in D(T(\omega))$ and $T(\omega,x) = y\}$. But

$$(1.2)\quad \left[\bigcap_{n=1}^{\infty} \bigcup_{S(x,r+\frac{1}{n})} \{\omega: x\in D(T(\omega)) \text{ and } T(\omega,x) \in S(y,\tfrac{1}{n})\}\right]^c$$

$$= \bigcup_{n=1}^{\infty} \bigcap_{S(x',r+\frac{1}{n})} \Big[\{\omega: x\in D(T(\omega)) \text{ and } T(\omega,x) \in S^c(y,\tfrac{1}{n})\} \cup \{\omega: x\notin D(T(\omega))\}\Big].$$

Now we invoke the separability of T. Let Z be a separant of T, N a null set as in Def. 1.4. Then

$$\bigcap_{x\in S(x',r+\frac{1}{n})} \{\omega: x\in D(T(\omega)) \text{ and } T(\omega,x) \in S^c(y,\tfrac{1}{n})\}$$

$$= (N \cap \bigcap_{S(x',r+\frac{1}{n})} \{\omega: x\in D(T(\omega)) \text{ and } T(\omega,x) \in S^c(y,\tfrac{1}{n})\} \cup$$

$$(N^c \cap \bigcap_{x\in S(x',r+\frac{1}{n})\cap Z} \{\omega: z\in D(T(\omega) \text{ and } T(\omega,z) \in S^c(y,\tfrac{1}{n})\}.$$

because of the separability of T. The first set is measurable because of the completeness of (Ω, A, μ), the second one by virtue of the randomness of T. As T is a random operator, $\{\omega : x \notin D(T(\omega))\} \in A$. Hence by (1.1) and (1.2) the result follows. □

Corollary 1.10: Let T be a bounded linear random operator on $\Omega \times X$ onto Y such that $T(\omega, \cdot)$ is invertible and its inverse $T^{-1}(\omega, \cdot)$ is bounded. Then T^{-1} is a random operator from $\Omega \times Y$ onto X.

Hanš ([28, p. 129] and [29, p. 192]) was the first one to prove the measurability of the inverse of a one-to-one and onto bounded random linear operator between two separable Banach spaces. The results of Nashed and Salehi [62] yield the measurability of inverses of (not necessarily bounded) linear operators if the inverse $T^{-1}(\omega, \cdot)$ is bounded, and furnish information on the measurability of the generalized inverse of a linear operator on a Hilbert space when either T or its generalized inverse $T^{\dagger}$ is bounded. In this paper we relax these assumptions and obtain a measurability result when neither T nor $T^{\dagger}$ is bounded (see Section II.4). This yields, in particular, a measurability result (Remark 2.9) for T^{-1}, when neither T nor T^{-1} is bounded, a result which hereto has not been available in the literature.

For a noninvertible operator T, the generalized inverse $T^{\dagger}$ can be defined as a suitable linear extension of the inverse of the restriction of T to a maximal subspace where it is invertible. For random operators this subspace will in general depend on ω. In this connection the following result [22] is useful in establishing measurability of generalized inverses.

Proposition 1.11: Let $T: \Omega \times X \to Y$ be a linear bounded random operator with closed range. Then the null space $N(T(\cdot))$ and the range $R(T(\cdot))$ are measurable. Suppose, moreover, that there is a measurable $M: \Omega \to 2^X$ such that for all $\omega \in \Omega$, $M(\omega)$ is a subspace of X with $X = N(T(\omega)) \oplus M(\omega)$ (a topological decomposition). Then $(T|M)^{-1}$: Gr $R(T(\cdot)) \to X$ defined by $(T|M)^{-1}(\omega)y :=$ the unique element x in $M(\omega)$ with $T(\omega)x = y$, is a linear bounded random operator with stochastic domain $R(T(\cdot))$.

5. Measurability of Adjoints of Random Linear Operators

Let T be a linear (not necessarily bounded) operator from $D \subset X$ into Y, $\overline{D} = X$, and let X^* and Y^* be the (topological) dual spaces of X and Y respectively. The adjoint operator to T is defined as the operator whose domain consists of all

$y^* \in Y^*$ for which there exists $x^* \in X^*$ such that $y^*[T(x)] = x^*(x)$ for all $x \in D$. In this case we define $T^*(y^*) := x^*$ and call T^* the <u>adjoint</u> operator. Thus

$$y^*[T(x)] = x^*(x) = T^*(y^*)(x). \tag{1.3}$$

It is well known that T^* is a closed linear operator on $D(T^*) \subseteq Y^*$ into X^*. However, $D(T^*)$ is in general not dense in Y^*. If $D(T^*)$ is dense in the weak-star topology in Y^* and Y is reflexive, then $D(T^*)$ is dense (in the norm topology) in Y^*. If T is bounded, then T^* is bounded and $||T^*|| = ||T||$. The following properties of the adjoint operator are also important:

(i) If T is from X to Y, S is from Y to Z, ST is densely defined in X, and S is bounded, then $(ST)^* = T^*S^*$.

(ii) If X and Y are reflexive and T from X to Y is closed and densely defined, then T^* is closed, densely defined, and $T^{**} = T$.

(iii) If X and Y are Hilbert spaces and if T is a densely defined closed linear operator from X to Y, then the transformations $(I + T^*T)^{-1}$ and $T(I + TT^*)^{-1}$ are defined everywhere and bounded (see, e.g., [68]). Moreover,

$$||(I + T^*T)^{-1}|| \le 1 \quad \text{and} \quad ||T(I + TT^*)^{-1}|| \le 1 .$$

If in addition $R(T)$ is closed, then for all $x \in N(T)^{\perp}$

$$||(I + T^*T)^{-1}x|| \le (1+m)^{-1}||x||,$$

where

$$m = \inf\{\langle T^*Tx,x\rangle: x \in N(T)^{\perp}, ||x|| = 1\} > 0.$$

Now, let T be a random operator from $\Omega \times X$ into Y. The adjoint operator T^* of T is the mapping defined by $x^* = T^*(\omega)y^*$ where

$$y^*(T(\omega)x) = x^*(x) \quad \text{for all } x \in D(T(\omega)). \tag{1.4}$$

Let T be a closed densely-defined (or a bounded) linear random operator from $\Omega \times X$ into Y. Then the Closed Range Theorem (see [75]) asserts that the following conditions are equivalent:

(i) $R(T(\omega))$ is closed in Y,

(ii) $R(T^*(\omega))$ is closed in X^*.

In particular, $T(\omega)$ is invertible if and only if $R(T^*(\omega)) = X^*$. Furthermore, if $T(\omega)$ is invertible, then $T^*(\omega)$ is invertible, and $(T^*(\omega))^{-1} = (T^{-1}(\omega))^*$.

Hans [28] has studied measurability of adjoints of <u>invertible bounded</u> linear random operators on separable Banach spaces, and proved the following result: If any of the operators T, T^{-1}, T^*, $(T^{-1})^*$ is random, then all four operators are random. Nashed and Salehi [62] proved a simple lemma concerning the adjoints of not necessarily bounded random linear operators in separable Hilbert spaces; the analogous result in Banach spaces is immediate:

<u>Proposition 1.12</u>: Let X and Y be separable Banach spaces and let T be a random linear operator from $\Omega \times X$ into Y such that $\bigcap_{\omega} D(T(\omega)) = D$ is dense in X. If X^* is separable (e.g., if X is reflexive), then for each $y^* \in \bigcap_{\omega} D(T^*(\omega))$, $T^*(\cdot)y^*$ is a generalized random variable. In particular, if T is a random bounded linear operator then so is T^*, and conversely.

<u>Proof</u>: Let $y^* \in \bigcap_{\omega} D(T^*(\omega))$, $x \in D$. Because of $(T^*(\omega)y^*)x = y^*(T(\omega)x)$, the randomness of T and the continuity of y^*, $(T^*(\cdot)y^*)x$ is measurable. As D is dense in X and $T^*(\omega)y^*$ is a <u>continuous</u> linear functional, $(T^*(\cdot)y^*)x$ is measurable for <u>all</u> $x \in X$. Now let $x^{**} \in X^{**}$. We identify X with its natural embedding into X^{**}. As X is dense in X^{**} in the weak-star topology on X^{**} and any ball in X^{**} is metrizable in the weak-star topology (cf. [14]), there exists a sequence $\{x_n\}$ in X such that $(T^*(\omega)y^*)x_n \to x^{**}(T^*(\omega)y^*)$ for all $\omega \in \Omega$. So $x^{**}(T^*(\cdot)y^*)$ is measurable. This together with [30, p. 74] implies that $T^*(\cdot)y^*$ is a random variable. □

In the case of Hilbert spaces, we identify as usual the dual space Y^* with the space Y itself and then the adjoint T^* is defined by the relation

$$(1.5) \qquad \langle T(\omega)x,y\rangle = \langle x,T^*(\omega)y\rangle \quad \text{for all } x \in D(T(\omega)),\ y \in D(T^*(\omega)).$$

For a closed densely defined (or bounded) linear random operator T between two Hilbert spaces X and Y, we have

$$(1.6) \qquad N(T^*(\omega))^{\perp} = \overline{R(T(\omega))}\ ,\ N(T(\omega))^{\perp} = \overline{R(T^*(\omega))}$$

and the following orthogonal decompositions are induced:

$$(1.7) \qquad X = N(T(\omega)) \oplus \overline{R(T^*(\omega))}$$

$$(1.8) \qquad Y = \overline{R(T(\omega))} \oplus N(T^*(\omega)).$$

The adjoint T^* and the operators T^*T and TT^* play an important role in the theory and approximation of generalized inverse operators in Hilbert spaces.

6. Measurability of Projectors

In many deterministic approximation schemes, projections onto subspaces or convex sets have to be carried out. As we shall see in Section III.2 one may have to choose the spaces or sets onto which one projects depending on the random parameter if one wants to adjust projection methods to the treatment of random operator equations. This motivates the investigation of the relationship between the measurability of set-(especially subspace-)valued maps and the randomness of associated projectors. Results of this type have been obtained in the finite-dimensional convex case in Rockafellar [69].

In this section we review some recent results obtained in [22] and [23] and refer to these papers for proofs and more general results, e.g., for set-valued projectors onto proximinal sets (see [70]).

Theorem 1.13: Let X be uniformly convex, $C\colon \Omega \to 2^X$ measurable such that for all $\omega \in \Omega$, $C(\omega)$ is convex. By $P_C\colon \Omega \times X \to X$ we denote the metric projector onto C (i.e., $P_C(\omega,\cdot)$ assigns each x the unique point in $C(\omega)$ closest to x). Then P_C is a continuous random operator.

Remark 1.14: As the fixed point set of $P_C(\omega,\cdot)$ is $C(\omega)$, the converse of Theorem 1.14 (namely the statement that randomness of P_C implies measurability of C) can be derived from results in random fixed point theory, e.g., Corollary 7 in [20].

As orthogonal projectors onto subspaces of a Hilbert space are metric projectors, we could derive from Theorem 1.13 a measurability result for orthogonal projectors. But for this special case we need more information than Theorem 1.13 gives us. Theorem 1.13 is essentially proved by a selection-theorem-approach, whereas we can use the representation of an orthogonal projector as a Fourier series to employ the limit-theorem-approach for proving the following result:

Theorem 1.15: Let X be a (separable) Hilbert space, $S\colon \Omega \to 2^X$ such that for all $\omega \in \Omega$, $S(\omega)$ is a (closed) subspace of X. $P\colon \Omega \times X \to X$ is defined such that for each $\omega \in \Omega$, $P(\omega,\cdot)$ is the orthogonal projector onto $S(\omega)$. Then the following statements are equivalent:

a) S is measurable.

b) P is a continuous random operator.

c) There exists a sequence of measurable functions, $g_1, g_2, \ldots : \Omega \to X$ such that for all $\omega \in \Omega, (g_1(\omega), g_2(\omega), \ldots)$ is an orthonormal basis of $S(\omega)$ (some $g_i(\omega)$ may be 0).

d) There exists a sequence of measurable functions $\phi_1, \phi_2, \ldots : \Omega \to X$ such that for all $\omega \in \Omega$, $\overline{\text{span}(\phi_1(\omega), \phi_2(\omega), \ldots)} = S(\omega)$.

Remark 1.16: At least the implications $a \Rightarrow b \Leftrightarrow c \Leftrightarrow d$ remain true if X is only a separable inner product space, if we assume that all $S(\omega)$ are complete (e.g., finite-dimensional).

In treating projection methods for random linear operator equations (see Section III.2) we will have to project onto images of subspaces under a random operator; for this we will use the following result:

Theorem 1.17: Let $S: \Omega \to 2^X$ be such that for each $\omega \in \Omega$, $S(\omega)$ is a closed subspace of X. We assume that S fulfills one of the four equivalent conditions of Theorem 1.15. Let $B: \Omega \times X \to Y$ be a bounded linear random operator (not necessarily with closed range). For all $\omega \in \Omega$, let $U(\omega) := B(\omega, S(\omega))$. Then $\omega \to \overline{U(\omega)}$ fulfills the conclusions of Theorem 1.15.

The results quoted in this section lay the foundation for "randomization" of projection methods (see Section III.2).

Further results about measurability of projectors onto subspaces (not necessarily in Hilbert spaces) will follow from a representation of projectors using generalized inverses and measurability results for generalized inverses (see Corollary 2.17).

II. GENERALIZED INVERSES OF RANDOM LINEAR OPERATORS

1. Generalized Inverses of Linear Operators in Banach and Hilbert Spaces

Let X and Y be Banach spaces and let T be a linear operator with dense domain $D(T)$ in X and range in Y. We assume that the null space of T, $N(T)$, has a topological complement, say M, and the closure of the range of T, $\overline{R(T)}$, has a topological complement, say S; i.e., there exist (continuous) projectors P, Q with $PX = N(T)$ and $QY = \overline{R(T)}$ and

$$X = N(T) \oplus M, \quad Y = \overline{R(T)} \oplus S.$$

Let $\tilde{T} := T|M$ be the restriction of T to M. Then $\tilde{T}: M \to R(T)$ is one-to-one and onto. We define the generalized inverse $T^{\dagger}$ of T as the linear extension of $\tilde{T}^{-1}$ to $R(T) \dot{+} S$ such that $T^{\dagger}S = \{0\}$. Note that $T^{\dagger}$ depends on the topological complements M and S, or equivalently on the projectors P and Q. Thus we write $T^{\dagger}_{M,S}$ or $T^{\dagger}_{P,Q}$ when this dependence is to be stressed, and omit these subscripts when dealing with fixed complements or projectors and there is no possibility of confusion. This is the <u>function-theoretic</u> definition of the generalized inverse, which we formalize as follows:

$$(2.1) \quad \begin{cases} D(T^{\dagger}_{P,Q}) := R(T) \dot{+} S \\ T^{\dagger}y := \tilde{T}^{-1}y \quad \text{for} \quad y \in R(T) \\ T^{\dagger}y := 0 \quad \text{for} \quad y \in S \\ T^{\dagger}(y_1+y_2) := T^{\dagger}y_1 + T^{\dagger}y_2 \quad \text{for} \quad y_1 \in R(T),\ y_2 \in S. \end{cases}$$

Note that equivalently $T^{\dagger}_{P,Q} = \tilde{T}^{-1}Q$ on $D(T^{\dagger}_{P,Q})$. Clearly $R(T^{\dagger}) = M$ and $N(T^{\dagger}) = S$.

It is easy to prove that $T^{\dagger}_{P,Q}$ can be characterized as the unique operator B which satisfies the following three equations:

$$(2.2) \quad \begin{cases} BTB = B \quad \text{on} \quad D(B) := R(T) \dot{+} S \\ TB = Q \quad \text{on} \quad D(B) \\ BT = I-P \quad \text{on} \quad D(T) \end{cases}$$

From (2.2) it follows that

$$(2.3) \qquad TBT = T \quad \text{on} \quad D(T).$$

Equations (2.2)-(2.3) are the Banach space analogue of the Moore-Penrose equations in Hilbert space (see (2.10)). For $y \in R(T)$, $u = T^{\dagger}y$ is a solution of the operator equation

$$(2.4) \qquad Tx = y$$

and it is the unique solution which lies in M. For $y \notin R(T)$, $u = T^{\dagger}y$ is a solution of the projectional equation (relative to S or Q):

$$(2.5) \qquad Tx = Qy$$

and it is the unique solution of (2.5) which lies in M. We call $T^{\dagger}_{P,Q}y$ the best approximate solution of (2.4) relative to Q.

The set U_y of all solutions of (2.5) is given by $U_y = T^{\dagger}y + N(T)$.

The existence of the generalized inverse of a linear operator T in Banach spaces hinges upon the existence of topological complements to $N(T)$ and $\overline{R(T)}$; the latter is not a trivial matter since a closed subspace of a Banach space need not have a topological complement. (For a recent survey on complemented subspaces see Kadets and Mityagin [35]). Even the assumption that T is a bounded or closed linear operator does not help in this connection. In the case of Hilbert spaces this difficulty does not arise since every closed subspace of a Hilbert space has a topological (and in particular an orthogonal) complement. Thus if T is either a bounded linear operator on a Hilbert space H_1 or is a closed densely defined linear operator with range in a Hilbert space H_2, then $N(T)$ is a closed subspace of H_1, and thus

$$H_1 = N(T) \oplus N(T)^{\perp} = N(T) \oplus \overline{R(T^*)},$$

and

$$H_2 = \overline{R(T)} \oplus R(T)^{\perp} = \overline{R(T)} \oplus N(T^*).$$

These orthogonal decompositions induce of course orthogonal (equivalently: self-adjoint) projectors P and Q. The generalized inverse relative to these orthogonal complements (equivalently: projectors) is called the <u>Moore-Penrose inverse</u> in Hilbert space. $T^{\dagger}$ can also be characterized in this case by the following extremal (least-squares) property:

$$(2.5)\quad \begin{cases} u := T^{\dagger}y \quad \text{for} \quad y \in R(T) + R(T)^{\perp} \quad \text{minimizes} \\ ||Tx-y|| \quad \text{over} \quad x \in D(T) \quad \text{and has smallest norm} \\ \text{among all other minimizers.} \end{cases}$$

<u>Lemma 2.1</u>: The following sets are identical:

$$(2.7)\qquad \{u: \inf_{x \in D(T)} ||Tx-y|| = ||Tu-y||\},$$

$$(2.8)\qquad \{u: \quad Tu = Qy\},$$

$$(2.9)\qquad \{u: \; T^*(Tu-y) = 0\}.$$

For a proof, see [58]. In the case of a bounded operator we may write (2.9) in the form $\{u: T^*Tu = T^*y\}$, so that the set of all least-squares solutions coincides with the set of all solutions of the normal equations $T^*Tu = T^*y$. The generalized inverse $T^{\dagger}$ in the case of Hilbert space can thus be characterized by the property that $T^{\dagger}y$ is the minimal-norm element in any of the sets in Lemma

2.1. In contrast, it should be noted that in case of Banach spaces while $T^{\dagger}y \in \{u : Tu = Qy\}$ it does not belong to any of the other two sets in Lemma 2.1. Also, $T^{\dagger}y$ is not necessarily of minimal norm among the elements of the set (2.8). For a complete analysis of extremal properties of $T^{\dagger}y$ in Banach spaces see [63].

It should be noted that $D(T^{\dagger})$ is dense in Y, and that $T^{\dagger}$ is in general an unbounded operator.

Lemma 2.2: Let T be a closed or bounded linear operator. Then the following statements are equivalent:

(a) $D(T^{\dagger}) = Y$,

(b) $R(T)$ is closed in Y,

(c) $T^{\dagger}$ is a bounded operator.

For a bounded linear operator on a Hilbert space, the equations (2.2) and (2.3) take the equivalent form - the Moore-Penrose equations

(2.10a) $TT^{\dagger}T = T$

(2.10b) $T^{\dagger}TT^{\dagger} = T^{\dagger}$

(2.10c) $(TT^{\dagger})^* = TT^{\dagger}$

(2.10d) $(T^{\dagger}T)^* = T^{\dagger}T$

In this form there is no redundancy among these equations; i.e., the first equation in (2.10) does not follow from the remaining equations.

For certain applications it suffices to consider generalized inverses which only satisfy a subset of the relations in (2.10); in such cases, one does not of course get a unique generalized inverse. Of particular interest are generalized inverses which satisfy one of the following conditions:

(i) (2.10a) and (2.10d)

(ii) (2.10a) and (2.10c).

These partial inverses provide minimal-norm solutions of solvable equations and least-squares solutions of inconsistent equations, respectively. See [5], [63].

We conclude this section by recalling the definition of inner and outer inverse, whose measurability in the case of random operators will be studied in Section II.6. Let X and Y be

Banach spaces and let $T:X \to Y$ be a bounded linear operator. An <u>inner inverse</u> for T is a linear operator B from Y to X such that

$$TBT = T . \tag{2.11}$$

While every linear operator has an inner inverse, not every bounded linear operator has a <u>bounded</u> inner inverse. $T \in L(X,Y)$ has a bounded inverse if and only if $N(T)$ and $R(T)$ are closed complemented subspaces of X and Y respectively.

A linear operator B from Y to X is called an <u>outer inverse</u> for T if

$$BTB = B . \tag{2.12}$$

Clearly the zero operator is always a (bounded) outer inverse. Every operator has a nonzero outer inverse, but not necessarily a bounded nonzero outer inverse. For properties of inner and outer inverses, see [60], [63].

2. Series and Integral Representations of Generalized Inverses of Linear Operators

Various series and integral representations for generalized inverses of bounded or closed linear operators between Hilbert spaces have been obtained in the literature. (See [25],[58],[54],[6], [44].) The spectral calculus of self-adjoint operators plays a key role in the derivation of most of these representations. A characteristic feature of these representations is that they only hold <u>pointwise</u> in the case of an operator with nonclosed range, whereas they hold uniformly if the operator has a closed range.

Series and integral representations can of course be used to provide approximations to the generalized inverse. In the case of random operators, they play another important role: they can be used together with the limit theorem approach to establish measurability of the generalized inverse. For Hilbert spaces, Nashed and Salehi [62] used, for example, a series representation to establish the measurability of the generalized inverse of a bounded operator, and an integral representation to prove, under some mild restrictions, the randomness of the generalized inverse of a closed operator with closed range. In Sections 4 and 5 we shall use series representations in proving several results on measurability of generalized inverses of random operators in Hilbert and <u>Banach</u> spaces. The following representations will be used:

Lemma 2.3: Let X and Y be Hilbert spaces and T a closed densely defined linear operator from X to Y. Then:

(a) For each $\alpha > 0$, the operator $(T^*T+\alpha I)^{-1}T^*$ has a continuous extension T_α to all of Y; for all $y \in D(T^\dagger)$, $\lim_{\alpha\to 0} T_\alpha y = T^\dagger y$.

(b) For every positive integer k, the operator $(I+T^*T)^{-k}T^*$ has a continuous extension B_k to all of Y; for all $y \in D(T^\dagger)$,

$$\sum_{k=1}^{\infty} B_k y = T^\dagger y.$$

Lemma 2.4: Let T be a bounded linear operator on a Banach space X into itself. Suppose that $\{(I-T^2)^n\}$ converges in the uniform operator topology to P, and $\{T(I-T^2)^n\}$ converges to the zero operator. Then the series $\sum_{k=1}^{\infty} T(I-T^2)^n$ converges to $T^\dagger_{P,I-P}$.

3. Tools for Establishing Measurability of Generalized Inverses of Linear Operators in Hilbert and Banach Spaces

All generalized inverses considered in this paper are defined on a dense domain. Therefore in the case of a random operator T, the domain $D(T^\dagger(\cdot))$ is measurable in the sense of Definition 1.1.

To establish measurability of the generalized inverse of a linear random operator T, Definition 1.2 suggests that we first prove the measurability of the set

$$\{\omega : y \in D(T^\dagger(\omega))\} \tag{2.13}$$

for each $y \in Y$. In the case of an operator with closed range, Lemma 2.2 shows that the set in (2.13) is always Ω, and thus to establish measurability of $T^\dagger$ it suffices to use a convergent approximation scheme in the spirit of the limit theorem approach (Section I.3). For the case of operators with nonclosed range, measurability problems are more difficult since we have to establish the measurability of the set in (2.13), which is not implied by the measurability of $D(T^\dagger(\cdot))$.

Lemma 2.5: Let X be a reflexive separable Banach space. Let T be a linear separable random operator from $\Omega \times X$ into Y with dense domain (i.e., for each $\omega \in \Omega$, $\overline{D(T(\omega))} = X$). Assume that for all $\omega \in \Omega$,

$$X = N(T(\omega)) \oplus M(\omega),\ Y = \overline{R(T(\omega))} \oplus S(\omega)$$

and that the projector $Q(\cdot)$ of Y onto $\overline{R(T(\cdot))}$ along $S(\cdot)$ is measurable. If T is either bounded or closed, then the set $\{\omega: y \in D(T^{\dagger}(\omega))\} \in A$ for all $y \in Y$.

Proof:

$$\{\omega: y \in D(T^{\dagger}(\omega))\} = \{\omega: y - Q(\omega)y = T(\omega)x \text{ for some } x \in X\}$$
$$= \bigcup_{x \in X} \{\omega: x \in D(T(\omega)) \text{ and } y - Q(\omega)y = T(\omega)x\}$$
$$= \bigcup_{N} \bigcup_{||x|| \le N} \{\omega: x \in D(T(\omega)) \text{ and } y - Q(\omega)y = T(\omega)x\}.$$

Thus it is enough to show that $\bigcup_{||x|| \le N} \{\omega: x \in D(T(\omega))$ and $y - Q(\omega)y = T(\omega)x\}$ is measurable. Let $S(y, \frac{1}{n}) := \{z: ||z-y|| < \frac{1}{n}\}$.

We assert that

$$\bigcup_{||x|| \le N} \{\omega: x \in D(T(\omega)) \text{ and } y - Q(\omega)y = T(\omega)x\}$$

$$(2.14) \quad = \bigcap_{n=1}^{\infty} \bigcup_{||x|| < N + \frac{1}{n}} \{\omega: x \in D(T(\omega)) \text{ and } y - Q(\omega)y \in S(y, \tfrac{1}{n})\}.$$

It suffices to show that the right-hand side is contained in the left-hand side, since the other inclusion is obvious. Let ω_o be in the right-hand side. Then for each $n \ge 1$, there exists some $x_n \in D(T(\omega_o)), ||x_n|| < N + \frac{1}{n}$, such that $T(\omega_o)x_n + Q(\omega_o)y \in S(y, \frac{1}{n})$. Since the sequence $\{x_n\}$ is bounded it follows from the reflexivity of X that there exists a subsequence $\{x_{n_k}\}$, $||x_{n_k}|| < N + \frac{1}{n_k}$, and x such that $x_{n_k} \to x$ weakly. Moreover, $||x|| < N$. From $T(\omega_o)x_n + Q(\omega_o)y \in S(y, \frac{1}{n})$, we obtain

$$T(\omega_o)x_{n_k} + Q(\omega_o)y \to y \text{ strongly.}$$

Now the graph of T is a closed and convex subset of $X \times Y$ and hence weakly closed. Therefore $x \in D(T(\omega_o))$ and $T(\omega_o)x + Q(\omega)y = y$. Thus

$$\omega_o \in \bigcup_{||x|| \le N} \{\omega: T(\omega)x + Q(\omega)y = y\}.$$

This proves (2.14). Taking complements in (2.14) we get

$$\left[\bigcup_{||x|| \le N} \{\omega: x \in D(T(\omega)) \text{ and } y - Q(\omega)y = T(\omega)x\}\right]^{c}$$

$$= \bigcup_{n=1}^{\infty} \bigcap_{||x||<N+\frac{1}{n}} \Big[\{\omega : x \in D(T(\omega)) \text{ and } T(\omega)x + Q(\omega)y \in S^c(y,\tfrac{1}{n})\} \cup$$

$$\{\omega : x \notin D(T(\omega))\}\Big]. \tag{2.15}$$

Since T is separable and Q is a random operator, it follows that for a fixed y, $Q(\cdot)y + T(\cdot)$ is a separable random operator from $\Omega \times X$ into Y. This, together with the completeness of (Ω, A, μ), implies that

$$\bigcap_{||x||<N+\frac{1}{n}} \{\omega : x \in D(T(\omega)) \text{ and } Q(\omega)y + T(\omega)x \in S^c(y,\tfrac{1}{n})\} \in A .$$

Since T is a random operator, $\{\omega : x \notin D(T(\omega))\} \in A$. Therefore by (2.14) and (2.15), the set $\{\omega : y \in D(T^\dagger(\omega))\}$ is measurable. □

Corollary 2.6: Let X and Y be Hilbert spaces and let T be a separable linear random operator from $\Omega \times X$ into Y with dense domain. If T is either bounded or closed, then the set $\{\omega : y \in D(T^\dagger(\omega))\} \in A$ for all $y \in Y$, where $T^\dagger$ is the Moore-Penrose inverse of the random operator T.

Proof: The corollary follows from Lemma 2.5, Theorem 1.9, and Theorem 1.15.

4. Measurability of Generalized Inverses of Random Linear Operators Between Hilbert Spaces

Throughout this section, we let X and Y be separable Hilbert spaces. Questions of measurability of $T^\dagger$ for a random linear operator T between Hilbert spaces were first investigated by Nashed and Salehi [62], who established the following results: (i) If T is bounded, then $T^\dagger$ is a random operator. (ii) If T is a closed operator with dense domain and if $R(T)$ is closed, then $T^\dagger$ is a random operator under some mild restrictions on the domains of $T(\omega)$ and $T^*(\omega)$. Thus the results of [62] provide information on the measurability of the generalized inverse in Hilbert space provided that either T or $T^\dagger$ is bounded. In this section we prove a measurability result for the generalized inverse where neither T nor $T^\dagger$ is necessarily bounded.

Theorem 2.7: Let T be a closed linear random operator from $\Omega \times X$ into Y such that the following conditions are satisfied:

(i) $\bigcap_{\omega} D(T(\omega)) =: D$ is dense in X;

(ii) $\bigcap_{\omega} D(T^*(\omega)) =: D^*$ is dense in Y;

(iii) $T^*(\cdot)T(\cdot)$ is separable.

Then $T^\dagger$ is a random operator from $\Omega \times Y$ into X.

Proof: Let n be a positive integer and let $\alpha = n^{-1}$. Then $T^*T + \alpha I$ is separable (by assumption (iii)) and has a bounded inverse on all of X for each ω (see Riesz-Nagy [68]). It follows from Theorem 1.9 that $(T^*T + \alpha I)^{-1}$ is a random operator. By Proposition 1.12 (here assumption (i) is used) and Lemma 1.3, $(T^*T + \alpha I)^{-1}T^*$ is a random operator on $\Omega \times D^*$. Because of the continuity of $(T^*T + \alpha I)^{-1}T^*$ (see Lemma 2.3) and assumption (ii), we can extend this operator to a bounded random operator T_α on $\Omega \times Y$. For each $\omega \in \Omega$, $T_\alpha(\omega)y \to T^\dagger(\omega)y$ for $y \in D(T^\dagger(\omega))$ by Lemma 2.3. To complete the proof, we apply Lemma 1.7 to the restriction of $T_\alpha(\cdot)y$ to the measurable set $\{\omega: y \in D(T^\dagger(\omega))\}$ (see Corollary 2.6). □

As a Corollary of Theorem 2.7 we obtain Theorem 2.2 of [62]:

Corollary 2.8: Let T be a bounded linear random operator on $\Omega \times X$ into Y. Then $T^\dagger$ is a random operator from $\Omega \times Y$ into X.

Proof: Since T^* is bounded, assumptions (i) and (ii) of Theorem 2.7 are satisfied. Assumption (iii) follows from Lemma 1.6. □

Remark 2.9: If T is one-to-one in Theorem 2.7 or Corollary 2.8, we obviously get results on measurability of T^{-1}. See also the remarks following Corollary 1.10.

Remark 2.10: In Corollary 2.8 we did not assume that T has closed range. An important class of linear operators with non-closed range are compact operators with infinite-dimensional range, e.g., Fredholm integral operators with non-degenerate kernel. The generalized inverse of a compact operator has a series representation in terms of a singular system. Therefore it is of interest to determine if singular systems are measurable, because this provides us with a stochastic version of an approximation scheme which is very useful in the deterministic setting.

Let $K:\Omega\times X \to Y$ be a compact linear random operator. Then the operators K^*K and KK^* are compact, self-adjoint, and non-negative. The operator KK^* has a finite or countably infinite number of nonzero eigenvalues λ_n, in the latter case $\lambda_n \to 0$ as

$n \to \infty$. Each nonzero eigenvalue is of finite multiplicity and the λ_n's can be linearly ordered: $\lambda_1 > \lambda_2 > \ldots > \lambda_n > \ldots > 0$, with mutually orthogonal eigenspaces $E_1, E_2, \ldots, E_n, \ldots$ where $\dim E_n =: m(n)$ is the multiplicity of λ_n. Let $\sigma(KK^*) = \{\lambda_1, \lambda_2, \ldots, \lambda_n, \ldots\}$ denote the nonzero spectrum of KK^*. It follows easily that $\sigma(KK^*) = \sigma(K^*K)$ and that the multiplicities of the eigenvalues are the same in both spectra. For each n, let $\{\psi_{n,1}, \ldots \psi_{n,m(n)}\}$ (resp. $\{\phi_{n,1}, \ldots \phi_{n,m(n)}\}$) be orthonormal bases of the eigenspace of K^*K (resp. KK^*) belonging to λ_n. It is well-known that we can choose the $\psi_{n,i}$ and $\phi_{n,i}$ in such a way that

$$\psi_{n,i} = \mu_n K^* \phi_{n,i}, \quad \phi_{n,i} = \mu_n K \psi_{n,i} \tag{2.16}$$

where $\mu_n = \lambda_n^{-1/2}$. We call $\{\phi_{n,i}, \psi_{n,i}; \mu_n\}$ a singular system for the compact operator K and μ_n a singular value of K. It is well-known (see [55]) that $y \in D(K^\dagger)$ if and only if $\sum_{n=1}^{\infty} \mu_n^2 \sum_{i=1}^{m(n)} |\langle y, \phi_{n,i}\rangle|^2 < \infty$ and that for $y \in D(K^\dagger)$

$$K^\dagger y = \sum_{n=1}^{\infty} \mu_n \sum_{i=1}^{m(n)} \langle y, \phi_{n,i}\rangle \psi_{n,i}. \tag{2.17}$$

If the number of singular values is finite, say N, we set $\mu_n := 0$ for $n > N$. Normally, (2.17) is written as

$$K^\dagger y = \sum_{n=1}^{\infty} \mu_n \langle y, \phi_n\rangle \psi_n,$$

where it is understood that each μ_n appears as often as its multiplicity; for random operators, these multiplicities depend on ω, hence the notation in (2.17) is more convenient for random operators.

Although there are measurability results for the resolvent set of a random operator (cf. [7, pp. 81-85]), there seems to be no result in the literature on the randomness of eigenvalues and eigenvectors of a compact self-adjoint nonnegative random operator on infinite-dimensional Hilbert spaces. In [23] we prove:

Proposition 2.11: Let $T: \Omega \times X \to X$ be a compact linear non-negative self-adjoint random operator. For each $\omega \in \Omega$, let $\lambda_1(\omega) > \lambda_2(\omega) > \lambda_3(\omega) > \ldots$ be the non-zero eigenvalues of $T(\omega)$ with eigenspaces $E_1(\omega), E_2(\omega), E_3(\omega), \ldots$. (If $T(\omega_o)$ has only n eigenvalues, then $\lambda_k(\omega_o) := 0$ and $E_k(\omega_o) := \{0\}$ for $k > n$).

Then for each $i \in \mathbb{N}$, λ_i is measurable and E_i is a measureable set-valued map; furthermore there exists a sequence

$\phi_{i,1}, \phi_{i,2}, \phi_{i,3}, \ldots$ of <u>measurable</u> functions from Ω into X such that for all $\omega \in \Omega$, the non-zero elements of $\{\phi_{i,1}(\omega), \phi_{i,2}(\omega), \phi_{i,3}(\omega), \ldots\}$ form an orthonormal basis of $E_i(\omega)$. Using this result, Proposition 1.12, and (2.17), we obtain a different constructive proof of Corollary 2.8 for compact operators:

<u>Corollary 2.12</u>: Let $K: \Omega \times X \to Y$ be a compact linear random operator. Then $K^\dagger$ is a random operator.

5. Generalized Inverses of Random Bounded Linear Operators in Banach Space

As noted in Section II.1, the generalized inverse of a linear operator T between two Banach spaces X and Y exists if $N(T)$ and $\overline{R(T)}$ have topological complements in X and Y respectively. Different choices of complements (equivalently: projectors) induce different generalized inverses. In Hilbert spaces, the choice of orthogonal projectors is prominently distinguished among all possible projectors because of the least-squares property of the Moore-Penrose inverse. Unfortunately, no choice of projectors enjoys such a particular distinction in general Banach spaces. Since such projectors are rarely metric projectors, we cannot conclude their measurability from Theorem 1.13; in fact, since complements are not unique there will be in general projectors onto $N(T)$ and $\overline{R(T)}$ which are not random operators. Hence one either assumes that measurable projectors exist or imposes conditions on T that simultaneously imply the existence and measurability of (continuous) projectors. Both of these cases will be illustrated below.

Convergent approximation schemes for $T^\dagger$ in the case of Hilbert spaces were crucial for proving measurability. These schemes are based on the connection between $T^\dagger$ and the self-adjoint operator T^*T which manifests itself in the normal equation. Attempts have been made to develop analogous convergent approximation schemes for $T^\dagger$ in the case of Banach spaces by replacing T^*T by BT (see, e.g., [40]). However, the only <u>concrete</u> realization of the "pseudoadjoint" B which has been given so far is $B = T$ in the case when T maps a Banach space X into itself. Thus, satisfactory results on measurability of $T^\dagger$ can so far be obtained only for this case as long as the proof is based on the limit theorem approach.

The selection theorem approach allows us to prove a measurability result for $T^{\dagger}$ where T maps X into Y (see Theorem 2.16) for the case of closed range.

The proofs of the following results are given in Engl and Nashed [22]. The first theorem uses the series representation given in Lemma 2.4, which is given in [41].

Theorem 2.13: Let T be a bounded linear random operator on $\Omega \times X$ into X. Assume that for all $\omega \in \Omega$, $\{(I - T^2(\omega))^n\}$ converges in norm to $P(\omega)$ and $\{T(\omega)(I - T^2(\omega))^n\}$ converges to the zero operator. Then

(i) $P(\cdot)$ is a measurable projector, $R(P(\omega)) = N(T(\omega))$, $R(I-P(\omega)) = R(T(\omega))$.

(ii) $T^{\dagger}_{P,I-P}$ is a random operator.

It should be noted that (i) is equivalent to the topological decomposition

$$X = N(T(\omega)) \oplus R(T(\omega))$$

which implies that $R(T(\omega))$ is closed.

Theorem 2.14: Let X be reflexive and T a bounded linear random operator on $\Omega \times X$ into X. Assume that for all $\omega \in \Omega$, $X = N(T(\omega)) \oplus \overline{R(T(\omega))}$ and $\{(I-T^2(\omega))^n x\}$ is convergent for all $x \in X$. Then the generalized inverse $T^{\dagger}_{\overline{R(T)},N(T)}$ is a random operator.

Theorem 2.15: Let T be a bounded linear random operator on $\Omega \times X$ into X. Assume that for all $\omega \in \Omega$,

(i) $X = N(T(\omega)) \oplus R(T(\omega))$

(ii) $\sigma(T(\omega))$, the spectrum of $T(\omega)$, is contained in some closed disk of radius $r(\omega)$ whose center lies on $(0,\infty)$ and whose boundary passes through 0, which is an isolated point of the spectrum. Furthermore, we assume that there exists a measurable function $g: \Omega \to \mathbb{R}$ such that $r(\omega) \leq g(\omega)$ for all $\omega \in \Omega$ (this holds, in particular, if $\sup r(\omega) < \infty$). Then $T^{\dagger}_{R(T),N(T)}$ is a random operator.

The proofs of the preceding three theorems are based on the limit-theorem approach. The selection-theorem approach enables us to prove the following result:

Theorem 2.16: Let $T: \Omega \times X \to Y$ be a bounded linear random operator with closed range, $M: \Omega \to 2^X$, $S: \Omega \to 2^Y$ measurable such that for all $\omega \in \Omega$, $M(\omega)$ and $S(\omega)$ are subspaces of X, Y, respectively, with $X = N(T(\omega)) \oplus M(\omega)$ and $Y = R(T(\omega)) \oplus S(\omega)$. Then $T^{\dagger}_{M,S}$ is a random operator on $\Omega \times Y$.

Idea of the proof: We first use Proposition 1.11 to conclude that $(T|M)^{-1}$ is a random operator with stochastic domain $\text{Gr}\, R(T(\cdot))$. Now, for any linear random operator L, let $\gamma(L(\omega)) := \{(x,y) : x \in D(L(\omega)), L(\omega)x = y\}$ be the graph of $L(\omega)$ (not to be confused with the graph of a set-valued map, cf. Def. 1.1). Since $T^{\dagger}(\omega)_{M(\omega),N(\omega)}$ is the unique linear extension of $(T(\omega)|M(\omega))^{-1}$ (defined on $R(T(\omega))$) and 0 (on $S(\omega)$), we have $\gamma(T^{\dagger}(\omega)_{M(\omega),N(\omega)}) = \gamma((T(\omega)|M(\omega))^{-1}) + (S(\omega) \times \{0\})$. We use measure-theoretic methods, including selection theorems, to conclude from this identity the randomness of $T^{\dagger}_{M,N}$. For details see [22]. □

As a consequence we obtain a measurability result for projectors onto certain random subspaces of a separable Banach space:

Corollary 2.17: Let T, M and S be as in Theorem 2.16. For all $\omega \in \Omega$ let $P(\omega)$ be the linear projector onto $N(T(\omega))$ parallel to $M(\omega)$, $Q(\omega)$ the linear projector onto $R(T(\omega))$ parallel to $S(\omega)$. Then P and Q (and therefore also $I-P$ and $I-Q$) are random operators.

Proof follows from (2.2), Theorem 2.16 and Lemma 1.3. □

Remark 2.18: In view of Corollary 2.17 and Proposition 1.11 the measurability of M and P (resp. S and Q) are equivalent. So one could be tempted to prove the measurability of P and Q without using Theorem 2.16 and then to use that $T^{\dagger} = (T|M)^{-1}Q$ and $(T|M)^{-1}$ and Q are random operators. The problem here, however, is that we do not know of a result about the measurability of the composition of two random operators in the absence of the separability assumption in Lemma 1.3. So we would have to assume the existence of a countable set Z such that for all $\omega \in \Omega$, $\overline{R(T(\omega)) \cap Z} = R(T(\omega))$, which is not needed in Theorem 2.16.

Corollary 2.19: Let T, M and S be as in Theorem 2.16. For all $\omega \in \Omega$, let $Q(\omega)$ be the linear projector onto $R(T(\omega))$ parallel to $S(\omega)$. Let $y: \Omega \to Y$ be measurable. Then the best-approximate solution of $T(\omega)x = y(\omega)$ relative to $Q(\omega)$ (cf. II.1) is a measurable function of ω.

Proof follows from Theorem 2.16 and Lemma 1.3. □

Remark 2.20: Corollary 2.19 yields the measurability of one particular solution of the projectional equation $T(\omega)x = Q(\omega)y$. Using the fact that all its solutions differ from $T^{\dagger}y$ by an element of $N(T)$, one can show that in the situation of Corollary 2.19 there exists a countable dense set of measurable solutions of the projectional equation. For details see [22]; the result there reads as follows for the case of orthogonal projectors in Hilbert spaces:

Theorem 2.21: Let X, Y be separable Hilbert spaces, $T: \Omega \times X \to Y$ a bounded linear random operator, $y: \Omega \to Y$ measurable such that for all $\omega \in \Omega$, $y(\omega) \in D(T^{\dagger}(\omega))$. Then there is a countable set of measurable functions $x_1, x_2, x_3, \ldots: \Omega \to X$ such that for all $\omega \in \Omega$, $\{x_1(\omega), x_2(\omega), \ldots\}$ is a dense subset of the set of all least-squares solutions of $T(\omega)x = y(\omega)$.

6. Measurability of Outer Inverses

Whereas in Section 4 we were concerned with the measurability of the generalized inverse of a linear random operator on a Hilbert space (i.e., the measurability of the unique linear operator determined by the Moore-Penrose equations (2.10)), we now investigate the measurability of linear operators fulfilling only part of the Moore-Penrose equations, especially of bounded outer inverses (cf. (2.12)) of a bounded random operator between separable Banach spaces X and Y. We do this by regarding the equation which determines outer inverses, namely (2.12), as a nonlinear (indeed quadratic) random equation in the space $L(Y,X)$ of all bounded linear operators on Y with values in X. We use a selection theorem approach to show that if T is a random operator, then

$$ST(\omega)S = S \tag{2.18}$$

has random solutions. But unfortunately the space $L(Y,X)$, equipped with the topology generated by the operator norm ("uniform operator topology") is in general not separable, even if X and Y are. But separability is an essential requirement for the type of selection theorems we use (e.g., Theorem 1.8). So we have to equip $L(Y,X)$ with a different topology which fits our needs. We recall that the "strong topology" on $L(Y,X)$ is the locally convex

Hausdorff topology generated by the seminorms $S \to ||Sy||$ $(y \in Y,$ $S \in L(Y,X))$. This topology, which is also called (for obvious reasons) "topology of pointwise convergence", has the following useful properties (cf. [9],[13]):

Lemma 2.22: Let $r > 0$, $K_r(Y,X) := \{S \in L(Y,X) : ||S|| \le r\}$. $K_r(Y,X)$, equipped with the strong topology, is a separable complete metrizable space; the function

$$\circ : \begin{array}{ccc} K_r(Y,X) \times L(X,Y) & \to & L(X,X) \\ (S \quad , \quad T) & \to & S \circ T \end{array}$$

is continuous with respect to the strong topologies on $K_r(Y,X)$, $L(X,Y)$, $L(X,X)$ with the strong topology.

If we want to treat (2.18) as a random equation in $L(Y,X)$, we have to use a (seemingly) different, but according to Lemma 2.24 equivalent, notion of measurability:

Definition 2.23: $G: \Omega \to L(Y,X)$ is called "strongly measurable" if for all strongly open $D \subseteq L(Y,X)$, $G^{-1}(D) \in \mathcal{A}$.

Lemma 2.24: For $G: \Omega \to L(Y,X)$ let $L: \Omega \times Y \to X$ be defined by $L(\omega,y) := G(\omega)y$. Then L is a random operator if and only if G is strongly measurable.

The proof of this result and the next theorem, which is the main result of this section, can be found in [22]. We give only a rough outline of the proof:

Theorem 2.25: Let $T: \Omega \times X \to Y$ be a bounded linear random operator. Then there exists a countable set of bounded linear random operators $S_n: \Omega \times Y \to X$ such that for all $n \in \mathbb{N}$ and $\omega \in \Omega$,

$$S_n(\omega)T(\omega)S_n(\omega) = S_n(\omega).$$

Moreover, for all $\omega_o \in \Omega$ the following holds: If $L \in L(Y,X)$ is an arbitrary outer inverse of $T(\omega_o)$, then there exists a sequence $n_1, n_2, n_3 \dots$ in $\mathbb{N}$ such that for all $y \in Y$,

$$\lim_{i\to\infty} S_{n_i}(\omega_o)y = Ly.$$

Outline of the proof: Let for $m \in \mathbb{N}$

$$A_m(\omega) := \{S \in K_m(Y,X) : ST(\omega)S = S\}$$

($K_m(Y,X)$ is defined as in Lemma 2.22). One shows that for all $\omega \in \Omega$, $A_m(\omega)$ is nonvoid and strongly closed and that A_m is a measurable set-valued map from Ω into $K_m(Y,X)$ (equipped with the strong topology). Because of the properties of $K_m(Y,X)$ (Lemma 2.22) we can apply a selection theorem similar to Theorem 1.8 to conclude the existence of a sequence of strongly measurable functions $S_{m1}, S_{m2}, \ldots : \Omega \to K_m(Y,X)$ such that for each $\omega_o \in \Omega$, $\{S_{m1}(\omega_o), S_{m2}(\omega_o), \ldots\}$ is a strongly dense subset of $A_m(\omega_o)$. Because of the definition of $A_m(\omega)$, all S_{mi} are bounded outer inverses of T and can be identified with random operators because of Lemma 2.24. Union over $m \in \mathbb{N}$ and the observation that convergence in the strong topology is nothing but pointwise convergence complete the argument. □

Thus each bounded linear random operator has measurable bounded outer inverses, the measurable bounded outer inverses are even dense in the topology of pointwise convergence in the set of all bounded outer inverses for every fixed $\omega_o \in \Omega$.

A similar result can be proved for inner inverses in an analogous way. There is only one significant difference: In the proof of Theorem 2.25 we could conclude that each $A_m(\omega)$ is nonvoid since 0 is an outer inverse of any linear operator, but not an inner inverse. So one has to add assumptions which guarantee that each $T(\omega)$ has a bounded linear inner inverse of norm $\leq m$ (independent of ω) for some m.

In a similar way one can prove the existence of partial inverses which fulfill various combinations of the Moore-Penrose equations (2.10a)-(2.10d). If one of the equations (2.10c), (2.10d) is involved, one has to replace the strong topology by the weak operator topology, since the forming of the adjoint is weakly, but not strongly continuous ([13]).

III. ITERATIVE AND PROJECTION METHODS FOR LEAST-SQUARES SOLUTIONS OF LINEAR OPERATOR EQUATIONS IN HILBERT SPACE

Random analogues of various iterative methods (e.g., steepest descent, conjugate gradient, successive approximation) and projection methods (e.g., Galerkin method, Ritz method, least-squares method) can be transferred to the random situation verbatim, as far as convergence and error estimates for each fixed ω are concerned. The only new aspects are measurability and statistical properties of the approximants. In the case of projection methods, this also involves the choice of projection schemes in which the approximating subspaces depend on ω.

In this chapter we address these salient features of random operator equation approximation. Since these features are for the most part independent from the particular deterministic approximation scheme, we investigate only two methods: the method of steepest descent as an example of an iterative method, and a Galerkin-type projection scheme.

1. Iterative Methods

Let $T: \Omega \times X \to Y$ be a random bounded linear operator. We consider iterative methods for least-squares solutions of the operator equation:

$$T(\omega)x = y(\omega) \tag{3.1}$$

where $y: \Omega \to Y$ is a random variable such that $y(\omega) \in D(T^{\dagger}(\omega))$. Let $U_y(\omega)$ denote the set of all least-squares solutions of (3.1). Then $U_y(\omega) = T^{\dagger}(\omega)y(\omega) + N(T(\omega))$ is a closed linear manifold for all ω. Let $P_U : \Omega \times X \to X$ be the metric projector onto U_y. Then

$$P_U(x) = T^{\dagger}y + P_N x$$

where P_N is the orthogonal projector of X onto $N(T)$. Note that $P_U(z)$ is the unique element in U_y which minimizes $||x-z||$ over $x \in U_y$. We consider iterative methods for computing

$$P_U(\omega)z(\omega) , \tag{3.2}$$

where z is a given random variable. In particular, the best approximate solution $T^{\dagger}(\omega)y(\omega)$ can be computed by taking $z = 0$. Let

$$J(\omega,x) := \frac{1}{2} ||T(\omega)x-y(\omega)||^2. \tag{3.3}$$

The method of steepest descent (see, e.g., [57]) for minimizing $J(\omega,x)$ is defined by the sequence

$$x_{n+1}(\omega) = x_n(\omega) - \alpha_n \operatorname{grad}J(\omega,x_n(\omega)) \tag{3.4}$$

where α_n is chosen to minimize $J(\omega,x_{n+1}(\omega))$, which indicates that in general α_n depends also on ω. It follows easily that

$$r_n(\omega) := \operatorname{grad} J(\omega,x) = T^*(\omega)T(\omega)x - T^*(\omega)y(\omega) \tag{3.5}$$

and

$$\alpha_n(\omega) = \frac{||r_n(\omega)||^2}{||T(\omega)r_n(\omega)||^2} \tag{3.6}$$

These relations together with Lemma 1.3 and Proposition 1.12 imply that r_n is a random operator and α_n is a random variable. Thus the iteration scheme (3.4) generates random variables if the initial approximation x_o is a random variable. Together with convergence results for the deterministic case (see [57]) this proves the following:

Theorem 3.1: Let $x_o: \Omega \to X$ be measurable. Then the method of steepest descent defined by (3.4),(3.5) and (3.6) generates random approximants and converges to $P_U(\omega)x_o(\omega) = T^{\dagger}(\omega)y(\omega) + P_N(\omega)x_o(\omega)$.

Remark 3.2: Convergence rates for the method of steepest descent for deterministic operator equations yield immediately corresponding rates for random operator equations. If T has closed range, then

$$(3.7) \quad ||x_n(\omega)-P_U(\omega)x_o(\omega)|| \le \left(\frac{M(\omega)-m(\omega)}{M(\omega)+m(\omega)}\right)^n ||x_o(\omega)-P_U(\omega)x_o(\omega)||,$$

where for each $\omega \in \Omega$,

$$M(\omega) := \sup\{\langle T^*(\omega)T(\omega)x,x\rangle : x \in N(T(\omega))^{\perp}, ||x|| = 1\}$$

and

$$m(\omega) := \inf\{\langle T^*(\omega)T(\omega)x,x\rangle : x \in N(T(\omega))^{\perp}, ||x|| = 1\}.$$

Clearly $M(\omega) = ||T(\omega)||^2$. It is easy to show that $m(\omega) = ||T^{\dagger}(\omega)||^{-2}$. Thus M and m are random variables. The pseudo-condition number of $T(\omega)$ is defined by

$$\kappa(\omega) := ||T(\omega)||\cdot||T^{\dagger}(\omega)||.$$

Clearly for $T \neq 0$, $\kappa(\omega) \ge 1$. Since $\kappa(\omega)^2 = M(\omega)[m(\omega)]^{-1}$, it follows from (3.7) that rate of convergence of the method of steepest descent for random operator equations with closed range is at least geometric with ratio $\frac{\kappa^2(\omega)-1}{\kappa^2(\omega)+1}$. So if we have uniform bounds on $\kappa(\omega)$ or uniform estimates on the spectral bounds $M(\omega)$ and $m(\omega)$, then the method of steepest descent with $x_o = 0$ converges to $T^{\dagger}(\omega)y(\omega)$ uniformly in ω. Other modes of convergence, such as stochastic convergence, can be analyzed similarly.

For the case of nonclosed range, error estimates for the method of steepest descent can be found in [58],[50]. Convergence in this case is only pointwise and rate is of the order $\frac{1}{n}$.

Analogues of various iterative methods for generalized inverses and least-squares problems (such as those discussed in [54],[58]) can be formulated for random operator equations in a similar manner. As far as measurability is concerned, no new aspects arise.

In addition to iterative methods, series and integral representations have been used in Nashed-Salehi [62] and in our Theorems 2.7, 2.12-2.15 to prove measurability of $T^\dagger$. We conclude this section by presenting a series of random variables which converges to the generalized inverse of a closed linear random operator. Let T satisfy the assumptions of Theorem 2.7. Then $(I+T^*T)^{-k}T^*$ can be extended to a bounded random operator B_k defined on all of Y. For each $\omega \in \Omega$ and $y \in D(T^\dagger(\omega))$, $\sum_{k=1}^{\infty} B_k(\omega)y = T^\dagger(\omega)y$. Moreover, if T has closed range, then the convergence is in the uniform operator topology and the following error estimate holds:

$$||T^\dagger(\omega) - \sum_{k=1}^{n} B_k(\omega)|| \le (1 + ||T^\dagger(\omega)||^{-2})^{-n} ||T^\dagger(\omega)||.$$

The proof of the randomness of the operators B_k follows the same line as the proof of Theorem 2.7. In the case of a bounded random operator we conclude as in the proof of Corollary 2.8 that the assumptions about T are fulfilled. It should be mentioned that in the case of operators with closed range error estimates are also obtainable for the sequence used in the proof of Theorem 2.7, namely for $\alpha > 0$,

$$||(T^*T + \alpha I)^{-1}T^* - T^\dagger|| \le \alpha ||T^\dagger||^3.$$

For the case of nonclosed range, only pointwise error estimates are available; also since $T^\dagger$ is unbounded all approximation schemes mentioned so far are numerically unstable. For problems involving operators with nonclosed range one should use suitable regularization schemes and related methods for the resolution of ill-posed problems (see [55]).

2. Projection Methods

In this section we summarize some of the theorems obtained in Engl and Nashed [23], where proofs and further results can be found. Throughout this section let X and Y be separable Hilbert spaces.

There is an extensive literature about projection methods for deterministic equations; for aspects closely related to our presentation we quote Krasnoselski [42] and Petryshyn [67]. For projection methods for finding best approximate solutions we refer to Nashed [54].

One approach to the solution of a deterministic linear operator equation

$$(3.8) \qquad Tx = y \quad (x \in X \, , \, y \in Y)$$

uses a "projection scheme"

$$(3.9) \qquad (X_n, Y_n; \, Q_n),$$

where X_n, Y_n are suitable subspaces of X, Y and Q_n is the orthogonal projector onto Y_n. One then replaces (3.8) by

$$(3.10) \qquad T_n x_n = y_n \, ; (x_n \in X_n)$$

where $T_n = Q_n T$, $y_n = Q_n y$.

If the X_n and Y_n are chosen independently from each other, this scheme is called the "Galerkin-Petrov method".

A frequently used choice of the approximation scheme involves $Y_n := TX_n$. The projection method in this case is called "least-squares method" because it is equivalent to minimizing $||Tx_n - y||$ in X_n.

If T is not onto and one wants to find least-squares solutions of $Tx = y$, the projection schemes involve subspaces of the type T^*Y_n or T^*TX_n.

These examples indicate that in the case of <u>random</u> equations the projection schemes will in general depend on the random parameter ω. This motivates the use of "stochastic projection schemes":

<u>Definition 3.3</u>: A sequence of triples $(X_n, Y_n; \, Q_n)$ will be called a "<u>stochastic projection scheme</u>" if $X_n: \Omega \to 2^X$ and $Y_n: \Omega \to 2^Y$ are measurable and for all $n \in \mathbb{N}$ and $\omega \in \Omega$, $X_n(\omega)$ and $Y_n(\omega)$ are closed subspaces and Q_n is the orthogonal projector onto Y_n. If $T: \Omega \times X \to Y$ is a linear random operator, we define $T_n := Q_n T$.

Theorem 1.15 gives various ways of characterizing stochastic projection schemes. Especially we can conclude from Theorem 1.15 and Lemma 1.3 that if $(X_n, Y_n; \, Q_n)$ is a stochastic projection scheme and T is a bounded linear random operator, then Q_n and T_n are random operators. Now let $\phi_n: \Omega \to X$ and $\psi_n: \Omega \to Y$ be sequences of measurable functions such that for each $\omega \in \Omega$, $\{\phi_n(\omega)\}$ resp. $\{\psi_n(\omega)\}$ are linearly independent and complete (if

X or Y are finite-dimensional, we replace infinite sequences by finite ones, of course). We will use $\{\phi_n\}$ and $\{\psi_n\}$ in various ways to construct stochastic projection schemes for finding (least-squares) solutions of linear operator equations of first and second kind (i.e., in the case of non-closed resp. closed range). The common idea of the proofs of the results below is a combination of the selection-theorem approach and the limit-theorem approach. After verifying by use of Theorems 1.15 and 1.17 that the chosen projection scheme is indeed a stochastic projection scheme, one establishes the existence of random solutions of the approximate equations by a selection-theorem approach. The convergence follows from corresponding deterministic results. We list four typical results: Theorems 3.4 and 3.5 deal with the case of closed range, Theorems 3.6 and 3.7 with the case of nonclosed range. The first result in each case treats the case where the operator is one-to-one.

Theorem 3.4: Let $T: \Omega \times X \to Y$ be a bounded linear random operator such that for all $\omega \in \Omega$, $T(\omega)$ is one-to-one and onto, $y: \Omega \to Y$ measurable. Let for all $\omega \in \Omega$ and $n \in \mathbb{N}$, $X_n(\omega) := \operatorname{span}\{\phi_1(\omega),\dots,\phi_n(\omega)\}$, $Y_n(\omega) := \operatorname{span}\{\psi_1(\omega),\dots,\psi_n(\omega)\}$ with orthogonal projector $Q_n(\omega)$. Then $(X_n, Y_n; Q_n)$ is a stochastic projection scheme. Assume furthermore that for all $\omega \in \Omega$ there are $c(\omega) > 0$ and $N(\omega) \in \mathbb{N}$ such for all $n > N(\omega)$ and $x_n \in X_n(\omega)$,

$$(3.11) \qquad ||T_n(\omega,x_n)|| \geq c(\omega)\cdot ||x_n||.$$

Then there are measurable functions $x, x_1, x_2, x_3, \dots : \Omega \to X$ with the following properties:

(i) For all $n \in N$ and $\omega \in \Omega$, $x_n(\omega) \in X_n(\omega)$.

(ii) For all $\omega \in \Omega$ and $n \geq N(\omega)$,

$$T_n(\omega,x_n(\omega)) = Q_n(\omega,y(\omega)).$$

(iii) For all $\omega \in \Omega$, $x(\omega)$ is the unique solution of $T(\omega,x) = y(\omega)$.

(iv) For all $\omega \in \Omega$ and $n \in \mathbb{N}$, $x_n(\omega)$ is the orthogonal projection onto $X_n(\omega)$ of the best approximate solution of $Q_n(\omega)T(\omega)P_n(\omega)x = Q_n(\omega,y(\omega))$ in X.

(v) For all $\omega \in \Omega$ and $n \in \mathbb{N}$, $x_n(\omega)$ is the best approximate solution of $T_n(\omega,x) = Q_n(\omega,y(\omega))$ in $X_n(\omega)$.

(vi) For all $\omega \in \Omega$, $\{x_n(\omega)\} \to x(\omega)$.

Thus, each approximation $x_n(\omega)$ is the best approximate solution of $T_n(\omega,x) = Q_n(\omega,y(\omega))$ in $X_n(\omega)$. As soon as the "Polski condition" (3.11) is fulfilled for a specific ω, this best approximate solution is in fact a solution of the approximate equation (ii). Note that the index from which the Polski condition is fulfilled may depend on ω!

If we choose $\psi_n(\omega) := T(\omega, \phi_n(\omega))$, Theorem 3.4 gives us a stochastic version of the least-squares method.

Theorem 3.5: Let $T: \Omega \times X \to Y$ be a bounded linear random operator with closed range, $y: \Omega \to Y$ measurable. For all $\omega \in \Omega$ and $n \in \mathbb{N}$, let $X_n(\omega) := \text{span}\,\{T^*(\omega)\phi_1(\omega),\ldots,T^*(\omega)\phi_n(\omega)\}$, $Y_n(\omega) := T(\omega,X_n(\omega))$ with orthogonal projector $Q_n(\omega)$. Then $(X_n,Y_n;Q_n)$ is a stochastic projection scheme. For all $n \in \mathbb{N}$ and $\omega \in \Omega$, $T_n(\omega,x) = Q_n(\omega,y(\omega))$ has a unique solution $x_n(\omega)$ in $X_n(\omega)$. Each x_n is measurable and for all $\omega \in \Omega$, $\lim_{n\to\infty} x_n(\omega) = T^{\dagger}(\omega)y(\omega)$.

Now we turn to the case of non-closed range. It is well-known from the deterministic situation that one cannot expect a projection method to yield a sequence of approximate solutions that converges to a solution, but only a sequence of approximants such that the residual tends to zero (cf. [42]).

The next result is a stochastic least-squares method for equations of the first kind:

Theorems 3.6. Let $T: \Omega \times X \to Y$ be a bounded linear random operator such that for all $\omega \in \Omega$, $T(\omega)$ is one-to-one and $R(T(\omega))$ is dense in Y. Let $y: \Omega \to Y$ be a random variable. For all $\omega \in \Omega$, define $X_n(\omega) := \text{span}\{\phi_1(\omega),\ldots,\phi_n(\omega)\}$, $Y_n(\omega) := T(\omega,X_n(\omega))$ with orthogonal projector $Q_n(\omega)$. Then $(X_n,Y_n;Q_n)$ is a stochastic projection scheme. Assume that for all $z \in Y$ and $\omega \in \Omega$, $\lim_{n\to\infty} Q_n(\omega,z) = z$. Then for all $\omega \in \Omega$ and $n \in \mathbb{N}$, $T_n(\omega,x) = Q_n(\omega,y(\omega))$ has a unique solution $x_n(\omega)$ in $X_n(\omega)$. Each x_n is measurable and for all $\omega \in \Omega$, $\lim_{n\to\infty} ||T(\omega,x_n(\omega))-y(\omega)|| = 0$.

Finally we describe a projection method for solving a linear operator equation of the first kind with non-trivial null space:

Theorem 3.7. Let $T: \Omega \times X \to Y$ be a linear bounded random operator, $y: \Omega \to Y$ measurable. For all $\omega \in \Omega$ and $n \in \mathbb{N}$, let

$X_n(\omega) := \text{span}\ \{T^*(\omega)T(\omega)\phi_1(\omega),\ldots,T^*(\omega)T(\omega)\phi_n(\omega)\}$, $Y_n(\omega) :=$ $T(\omega,X_n(\omega))$ with orthogonal projector $Q_n(\omega)$, $Q(\omega)$ the orthogonal projector onto $\overline{R(T(\omega))}$. Then $(X_n, Y_n; Q_n)$ is a stochastic projection scheme. For all $\omega \in \Omega$ and $n \in \mathbb{N}$, $T_n(\omega,x) = Q_n(\omega,y(\omega))$ has a unique solution $x_n(\omega)$ in $X_n(\omega)$. Each x_n is measurable, for all $\omega \in \Omega$, $\lim_{n\to\infty}||T(\omega,x_n(\omega)) - Q(\omega,y(\omega))|| = 0$ and $\lim_{n\to\infty}||T(\omega,x_n(\omega)) - y(\omega)|| = \inf_{x\in X}||T(\omega,x) - y(\omega)||$.

IV. APPLICATIONS TO RANDOM INTEGRAL, DIFFERENTIAL, AND OPERATOR EQUATIONS

In this chapter we apply some of the results and techniques described in this paper to linear integral equations with random kernels, boundary- and eigen-value problems for differential equations with random coefficients, and nonlinear random operator equations. As a preamble we address two technical points which arise in the treatment of random integral and differential equations.

1. The Interplay Between Integral Operators and Their Generalized Inverses in the Spaces $L_2[0,1]$ and $C[0,1]$

A. Integral operators are frequently studied as acting either on $C[0,1]$ or on $L_2[0,1]$. In the framework of random integral operators, the need then arises to relate the concepts of measurability of the integral operators acting on these spaces and to characterize them in terms of the measurability of the kernels.

Lemma 4.1: Let $k:\Omega \times [0,1]^2 \to \mathbb{R}$ be such that for all $\omega \in \Omega$, $k(\omega, \cdot, \cdot)$ is jointly continuous at all (s,t) which do not lie on any of finitely many curves $t = \phi_i(s)$ (independent of ω) and separately continuous everywhere. For $x \in L_2[0,1]$, define

$$K(\omega,x) := \int_0^1 k(\omega, \cdot, t)\, x(t)dt. \tag{4.1}$$

Then the following statements are equivalent:

(i) $\underset{\sim}{K}$ is a random operator on $L_2[0,1]$ into itself.

(ii) $K := \underset{\sim}{K}|C[0,1]$, the restriction of $\underset{\sim}{K}$ to $C[0,1]$, is a random operator on $C[0,1]$ into itself.

(iii) The kernel k is a "random kernel", i.e., for all s and t in $[0,1]$, $k(\cdot, s, t)$ is measurable.

Proof: (i) $\Rightarrow$ (iii): Let (s_o, t_o) be chosen such that it does not lie on one of the curves $t = \phi_i(s)$. Then there exists an integer N such that k is jointly (and therefore uniformly) continuous on

$$[s_o - \frac{1}{N}, s_o + \frac{1}{N}] \times [t_o - \frac{1}{N}, t_o + \frac{1}{N}].$$

(Here and below expressions like $s_o - \frac{1}{N}$ and $s_o + \frac{1}{N}$ should be understood as $\max\{0, s_o - \frac{1}{N}\}$ and $\min\{1, s_o + \frac{1}{N}\}$, respectively.) For $s \in [0,1]$ and $n \geq N$, let

$$h_{s,n}: [0,1] \to R$$

$$t \to \begin{cases} b_{s,n} & \text{if } t \in [s - \frac{1}{n}, s + \frac{1}{n}] \\ 0 & \text{otherwise} \end{cases}$$

where $b_{s,n} := (\min\{s + \frac{1}{n}, 1\} - \max\{s - \frac{1}{n}, 0\})^{-1}$. Clearly, $h_{s,n}$ is in $L_2[0,1]$.

Let $\omega \in \Omega$ and $\varepsilon > 0$ be fixed, but arbitrary. Then

$$|\langle K(\omega)h_{t_o,n}, h_{s_o,n}\rangle - k(\omega, s_o, t_o)|$$

$$= \left| \int_{s_o-\frac{1}{n}}^{s_o+\frac{1}{n}} \int_{t_o-\frac{1}{n}}^{t_o+\frac{1}{n}} k(\omega,\sigma,\tau)h_{t_o,n}(\tau)d\tau h_{s_o,n}(\sigma)d\sigma - k(\omega,s_o,t_o)\right|$$

$$\leq b_{t_o,n} \cdot b_{s_o,n} \cdot \int_{s_o-\frac{1}{n}}^{s_o+\frac{1}{n}} \int_{t_o-\frac{1}{n}}^{t_o+\frac{1}{n}} |k(\omega,\sigma,\tau) - k(\omega,s_o,t_o)| d\tau d\sigma.$$

Because of the uniform continuity of $k(\omega,.,.)$ in

$$[s_o - \frac{1}{n}, s_o + \frac{1}{n}] \times [t_o - \frac{1}{n}, t_o + \frac{1}{n}],$$

the integrand can be bounded uniformly by ε if we choose n sufficiently large. This implies that

$$\lim_{n\to\infty} \langle K(\omega)h_{t_o,n}, h_{s_o,n}\rangle = k(\omega,s_o,t_o).$$

As K is a random operator, each $\langle K(\cdot)h_{t_o,n}, h_{s_o,n}\rangle$ is measurable, which implies the measurability of $k(\cdot,s_o,t_o)$. If (s_o,t_o) lies on one of the curves $t = \phi_i(s)$, then $k(\cdot,s_o,t_o)$ is measurable, since $k(\omega,s_o,t_o) = \lim_{t\to t_o} k(\omega,s_o,t)$ and $k(\cdot,s_o,t)$ is measurable for $0 < |t - t_o| < \varepsilon$ for ε sufficiently small by the preceding.

(iii) $\Rightarrow$ (i): k is jointly measurable (in all three variables). Let $x,y \in L_2[0,1]$. Then

$$\langle K(\cdot)x,y\rangle = \int_0^1 \int_0^1 k(\omega,s,t)\ x\ (t)dty(s)ds$$

is measurable according to a suitable form of Fubini's Theorem. This implies the measurability of $K(\cdot)x$ (see [7, p. 16]).

(ii) $\Longleftrightarrow$ (iii): [7, p. 154]. □

The most important application of Lemma 4.1 is to the case of kernels which are jointly continuous at all (s,t) with $s \neq t$ and separately continuous everywhere. This includes of course Green's functions and generalized Green's functions.

B. We consider the random integral equation of the second kind on $L_2[0,1]$

(4.2) $\qquad A(\omega,x) := x - K(\omega,x) = y(\omega)$

where $N(A(\omega,\cdot)) \neq \{0\}$ for some ω and where K is the operator defined in (4.1). We now regard K both as an operator on $L_2[0,1]$ and an operator $\tilde{K}$ on $C[0,1]$. Thus $\tilde{K} = K|C[0,1]$. Besides (4.2) consider in $C := C[0,1]$ the equation

(4.3) $\qquad \tilde{A}(\omega,x) = x - \tilde{K}(\omega,x) = y(\omega)$.

We assume that $K:\Omega \times L_2 \to L_2$ and $\tilde{K}:\Omega \times C \to C$. The generalized inverse of a bounded linear operator T on a Banach space X to a Banach space does not possess the least-squares property. Thus the least-squares property does not hold for integral equations of the second kind on the Banach space $C[0,1]$ with the best approximate solution taken in the sense of the uniform norm. However, we may study integral equations on C with the best approximate solution taken in the L_2-norm. The generalized inverse of $I-K$ in the sense of L_2 still yields a <u>continuous</u> best approximate solution, i.e., whereas $(I - K)^{\dagger}: L_2 \to L_2$, under mild conditions the restriction of $(I - K)^{\dagger}$ to C has its range in C, and $(I - K)^{\dagger}|C$ is in fact the generalized inverse of $\tilde{A}$ relative to the (restricted) orthogonal projectors. This interplay enables us to exploit the least-squares property of the generalized inverse in L_2, while treating the equation (4.3) in the space C. Following Moore and Nashed [51] the basic assumptions under which this interplay holds are:

(i) $\tilde{K}: C \to C$

(ii) $K: L_2 \to L_2$ (and hence $K^*: L_2 \to L_2$)

(iii) K and K^* map L_2 into C

(iv) For some $n > 0$, K^n is a compact operator on L_2; consequently (iv)a: $(I - K) L_2$ is closed, and (iv)b: $\dim N(I-K)$ and $\operatorname{codim} (I - K)L_2 = \dim ((I - K) L_2)^{\perp}$ are finite.

These assumptions are satisfied if $k(s,t)$ is continuous, or mildly singular, for example, $k(s,t) = \ell n|s - t|$ or $|s - t|^r$, $r > -1$.

Let $N: = N(A)$, $N^*: = N(A^*)$, $R: = AL_2$, $\tilde{R}: = \tilde{A}C$. By (iii), $N \subset C$, $N^* \subset C$, and

(4.4) $Ax = (I - K)x \in C \iff x \in C$, so $R \cap C = \tilde{R}$. By (ii) and (iv)a N and R are closed, and $R^{\perp} = N^*$. Hence we have

(4.5) $L_2 = N \oplus N^{\perp} = R \oplus N^*$ and the <u>orthogonal</u> decomposition of C:

(4.6) $C = N \oplus (N^{\perp} \cap C) = \tilde{R} \oplus N^*$.

Let Q, P, $I - P$ be the (continuous) orthogonal projectors of L_2 onto R, N and $N^{\perp}$, respectively. Then by the above

$$Qx \in C \iff x \in C \quad \text{and} \quad Px \in C \iff x \in C.$$

Let $A^{\dagger} = (I - K)^{\dagger}$ be the generalized inverse of A relative to the projectors P and Q, and let U_y be the set of all least-squares solutions of (4.2). Let K satisfy assumptions (ii), (iii) and (iv)a. Then

(4.7) $(I - K)^{\dagger} y \in C \iff y \in C$

and

(4.8) $y \in C \Rightarrow U_y \subset C$; $y \notin C \Rightarrow U_y \cap C = \phi$.

To apply the preceding results to $\tilde{A} = I - \tilde{K}$ in the space C, note that $\tilde{A} = A|C$, and let $A^{\#}: = A^{\dagger}|C$, $\tilde{P}: = P|C$, $\tilde{Q}: = Q|C$.

In view of (4.4) and (4.7), the restrictions of the Moore-Penrose inverse $A^{\dagger}$ to C imply

$$A^{\#}AA^{\#} = A^{\#}, \; A^{\#}A = I - P, \; AA^{\#} = Q.$$

Thus we have:

<u>Lemma 4.2</u>: Let K, $\tilde{K}$ satisfy assumptions (i), (ii), (iii) and (iv)a. Then $A^{\#} = A^{\dagger}|C$ is the generalized inverse in the <u>Banach space</u> C relative to the projectors $\tilde{P}: = P|C$ and $\tilde{Q}: = Q|C$.

Lemma 4.2 enables us to consider convergence properties of quadrature and collocation approximation for least-squares solutions of (4.3). The random analogue of Lemma 4.2 (where now K and $\tilde{K}$ are <u>random</u> integral operators on L_2, respectively C, into itself) obviously holds.

2. Random Resolvents and Pseudoresolvents; Explicit Representation of the Random Operator $(I - \lambda(\omega)\, K(\omega))^{\dagger}$

In this section we let X and Y be $L_2[0,1]$ and let $k(\omega,s,t)$ be measurable in ω and (jointly) continuous in (s,t). Let $K: \Omega \times X \to Y$ be the random Fredholm operator induced by the kernel $k(\omega,s,t)$ as in (4.1). Let λ be a real- or complex-valued measurable function on Ω. If $(I - \lambda K)^{-1}$ exists for all ω, then for each $\omega \in \Omega$,

$$(4.9) \qquad (I - \lambda(\omega)\, K(\omega))^{-1} = I + \lambda(\omega)\, \Gamma_{\lambda(\omega)}(\omega)$$

where Γ_{λ} is the usual resolvent operator. It is well-known that Γ_{λ} is a Fredholm integral operator. The kernel of $\Gamma_{\lambda(\omega)}(\omega)$ which we denote by $\gamma(\omega,s,t;\lambda(\omega))$ is called the resolvent kernel of $k(\omega,s,t)$. It is well known from the classical theory of Fredholm integral equations that

$$(4.10) \qquad \gamma(s,t;\lambda) - \frac{D(s,t;\lambda)}{d(\lambda)},$$

where,

$$(4.11) \qquad D(s,t;\lambda): = k(s,t) + \sum_{\lambda=1}^{\infty} \lambda^{n} D_{n}(s,t),$$

and

$$(4.12) \qquad d(\lambda): = 1 + \sum_{n=1}^{\infty} \lambda^{n} C_{n}.$$

The coefficients C_n, and the functions $D_n(s,t)$ can be determined recursively by

$$(4.13)\quad \begin{cases} C_1 = \int_0^1 k(s,s)ds,\ D_1(s,t) = C_1 k(s,t) + \int_0^1 k(s,u)k(u,t)du \\ C_n = \int_0^1 D_{n-1}(s,s)ds, \\ D_n(s,t) = C_n k(s,t) + \int_0^1 k(s,u)D_{n-1}(u,t)du \end{cases}$$

Both series in (4.11) and (4.12) converge for all λ, and $d(\lambda) = 0$ if and only if λ is a characteristic value of K.

Proposition 4.3. If $(I - \lambda(\omega)\ K(\omega))^{-1}$ exists for all ω, then the resolvent operator Γ_λ is a random operator from $\Omega \times Y$ into X and the function $\omega \to \gamma(\omega,\cdot,\cdot;\lambda(\omega))$ is measurable from Ω into $L_2[0,1]^2$.

Proof: The randomness of Γ_λ follows from Corollary 1.10 and the fact that $\Gamma_0 = K$ (see (4.11) and (4.12)). In order to prove measurability of γ we first note that by properties of iterated Lebesgue integrals the coefficients C_n and D_n are measurable functions of ω, and then apply Lemma 1.7 to the expression for γ given by (4.10). □

We now remove the assumption that $I - \lambda(\omega)K(\omega)$ is invertible for all ω; i.e., we allow $\lambda(\omega)$ to be a characteristic value of $K(\omega)$ for some ω. From the Riesz-Schauder theory (see, e.g., [72]) we know that for each ω

$$\dim N(I - \overline{\lambda}(\omega)\ K^*(\omega)) = \dim N(I - \lambda(\omega)\ K(\omega)) = n(\omega) < \infty.$$

Since $I - \overline{\lambda}K^*$ is a random operator, we conclude from Proposition 1.11 that $N(I - \overline{\lambda}(\cdot)K^*(\cdot))$ and $N(I - \lambda(\cdot)K(\cdot))$ are both measurable. Thus by Theorem 1.5 there exist sequences $\{\phi_1, \phi_2, \ldots\}$ and $\{\Psi_1, \Psi_2, \ldots\}$ of measurable functions such that for each ω exactly $n(\omega)$ elements in each sequence are nonzero and these elements (whose indices depend on ω) form orthonormal bases for $N(I - \lambda(\omega)\ K(\omega))$ and $N(I - \overline{\lambda}(\omega)\ K^*(\omega))$, respectively.

We now construct a random analogue to the Hurwitz pseudoresolvent, which extends the theory of resolvent operators to the case when $I - \lambda K$ is not invertible. For a development of the Hurwitz pseudoresolvent in the framework of generalized inverses, see [55].

Define

(4.14) $\tilde{k}(\omega,s,t) := k(\omega,s,t) - \frac{1}{\lambda(\omega)} \sum_{i=1}^{\infty} \Psi_i(\omega,s)\overline{\phi_i(\omega,t)}$,

where for each ω, the sum in (4.14) consists of finitely many nonzero terms. Furthermore, if $\lambda(\omega)$ is a regular value (in particular if $\lambda(\omega) = 0$), then $\tilde{k}(\omega,s,t) = k(\omega,s,t)$. Since the ϕ_i's and Ψ_i's are continuous it follows that $\tilde{k}$ is a continuous random kernel (in the sense of Lemma 4.1). Let $\tilde{K}$ be the integral operator induced by $\tilde{k}$, which by Lemma 4.1 is a random operator. Then $I - \lambda\tilde{K}$ is invertible and therefore has a resolvent $\tilde{\Gamma}_\lambda$ with resolvent kernel $\tilde{\gamma}$:

$$(I - \lambda\tilde{K})^{-1} = I + \lambda\tilde{\Gamma}_\lambda.$$

According to Hurwitz, an L_2 - kernel $\gamma(s,t;\lambda)$ is called a pseudo-revolent kernel of $k(s,t)$ if for each $y \in R(I - \lambda K)$, the function

$$y(s) + \lambda \int_0^1 \gamma(s,t;\lambda)y(t)dt$$

is a solution of $x - \lambda Kx = y$. The integral operator Γ_λ induced by γ is called a pseudoresolvent. Note that Γ_λ is a pseudo-resolvent if and only if $I + \lambda\Gamma_\lambda$ is a bounded inner inverse (see (2.11)) of $I - \lambda K$. It follows from the deterministic theory that $\tilde{\gamma}$ is a pseudoresolvent kernel. The measurability of $\tilde{\gamma}$ and $\tilde{\Gamma}$ follows from their construction and Lemma 4.1. Since in the deterministic case, every pseudoresolvent kernel can be written in the form

$$\tilde{\gamma}(s,t) + \sum_{i,j=1}^{n} c_{ij}\, \phi_i(t)\, \overline{\Psi_j(s)},$$

we have proved the following proposition:

Proposition 4.4. Let $I - \lambda(\omega)\, K(\omega)$ be not necessarily invertible for some ω. Then

(a) the function $\omega \to \tilde{\gamma}(\omega,\cdot,\cdot;\lambda(\omega))$ is measurable and for each ω, $\tilde{\gamma}(\omega,s,t,\lambda(\omega))$ is a pseudoresolvent kernel. Every measurable pseudoresolvent kernel is of the form

(4.15) $$\tilde{\gamma}(\omega,s,t) + \sum_{i,j=1}^{\infty} c_{ij}(\omega)\, \phi_i(\omega,t)\, \overline{\Psi_j(\omega,s)}$$

where the c_{ij}'s are arbitrary measurable functions.

(b) $\tilde{\Gamma}$ is a random operator and for each ω, $\tilde{\Gamma}(\omega)$ is a pseudoresolvent operator.

Remark 4.5: Obviously the representation (4.15) gives rise to a characterization of all random pseudoresolvent operators and all random bounded inner inverses of $I - \lambda K$. Note that $I + \lambda\tilde{\Gamma}$ is not an outer inverse (and hence not a generalized inverse) unless $I - \lambda K$ is invertible. Finally we remark that unlike the approach of II.2, we were in this section able to obtain measurability results for inner inverses of a special class of linear random operators without the assumptions discussed following Theorem 2.25.

3. Random Generalized Green's Function for an nth Order Random Linear Differential Operator

The generalized Green's function for a linear differential operator has been thoroughly investigated in the literature and its characterization in terms of the generalized inverse of the differential operator has been considered by several authors in recent years. For a biographical account and a sketch of the development of generalized inverses of differential operators and generalized Green's matrices, the reader is referred to [61; pp. 804-810] and the references cited therein.

There are several definitions of random or stochastic Green's functions in the literature (see [4], [7] and the references cited therein). These definitions, however, do not seem to provide the random analogue of the interrelations that exist in the deterministic case between the generalized Green's function and generalized inverse operators. The purpose of this section is to present a new characterization of a random generalized Green's function $G(t,s)$ for an nth order random linear differential operator L, which is determined by a formal random differential operator T and linearly independent boundary conditions $B_i(f) = 0$, $i = 1, \ldots, k$. This will provide a random analogue for the characterization of the generalized Green's function given in [48]. To motivate the presentation, we first review and recast in operator-theoretic terms the notions of Green's function and generalized Green's function for a formal second order linear differential operator.

The Green's function $G(t,s)$ for the (formally) selfadjoint operator

$$(4.16) \qquad Tx: = -\frac{d}{dt}\left(p(t)\frac{dx}{dt}\right) + q(t)x, \quad 0 < t < 1$$

with the prescribed boundary conditions

(4.17) $\quad a_1x(0) + b_1x'(0) = 0 \quad$ and $\quad a_2x(1) + b_2x'(1) = 0,$

where $|a_i| + |b_i| \neq 0,\ i = 1,2,$ can be constructed if and only if the equation $Tx = 0$ has no nontrivial solutions satisfying both boundary conditions (4.17), i.e., when zero is not an eigenvalue. In this case it is possible to find linearly independent functions u and v which satisfy the differential equation $Tu = 0$ and the first and second boundary conditions respectively. The Green's function is then explicitly constructed:

$$G(t,s) = \begin{cases} \dfrac{u(t)v(s)}{C} & 0 < t \leq s < 1 \\ \dfrac{u(s)v(t)}{C} & 0 < s \leq t < 1, \end{cases}$$

where $C = p(t)W[u(t),v(t)]$ is constant (W is the Wronskian of u and v). From the operator viewpoint, the Green's function is the kernel of the integral representation of the inverse of T:

$$T^{-1}y =: Gy = \int_0^1 G(\cdot,s)y(s)ds.$$

Thus if zero is not an eigenvalue of T, then T^{-1} exists on $R(T)$ and there exists a one-to-one correspondence between $BC^2[0,1]$, the space of all functions in $C^2[0,1]$ satisfying the boundary conditions, and the space $C[0,1]$.

If zero is an eigenvalue of T, the Green's function cannot be constructed. The equation $Tx = y$ has no solution unless y is orthogonal to the (one-dimensional) eigenfunction manifold M_o associated with $\lambda = 0$. However if we restrict ourselves to the orthogonal complements of M_o in $C[0,1]$ and $BC^2[0,1]$, denoted by $\tilde{C}$ and $B\tilde{C}^2$ respectively, then there is a one-to-one correspondence between these two subspaces. We denote $\tilde{T} := T|\tilde{C}$ and let $H = \tilde{T}^{-1}$. Then H is an integral operator (a Green's operator) whose kernel $K(t,s)$ is called the generalized Green's function. Explicit construction of $K(t,s)$ is of course given in standard books; however, the operator-theoretic interpretation is not stressed. It should be noted that in this setting we have the <u>topological</u> decompositions:

(4.18) $\quad BC^2[0,1] = M_o \oplus B\tilde{C}^2 = N(T) \oplus B\tilde{C}^2$

(4.19) $\quad C[0,1] = M_o \oplus \tilde{C} = N(T) \oplus \tilde{C} = N(T) \oplus R(T)$

and that the linear extension of H to all of $C[0,1]$ so that the extended operator is zero on $N(T)$ is the generalized inverse of T relative to these decompositions.

After this motivation, we now proceed directly to developing the random analogue of the generalized Green's function for boundary - value problems for an nth order linear differential operator. Consider the formal random differential operator

(4.20) $$T(\omega,x)(t) := \sum_{i=1}^{n} a_i(\omega,t) \frac{d^i}{dt^i} x(t)$$

where the coefficients a_i are measurable in ω, and $C^\infty[0,1]$ in t, $a_n(\omega,t) \neq 0$, and (for each ω) the linearly independent boundary operators

(4.21) $$B_i(\omega,f) := \sum_{j=0}^{n-1} \{\alpha_{ij}(\omega)\ f^{(j)}(a) + \beta_{ij}(\omega) f^{(j)}(b)\},\ i = 1,\ldots,k,$$

where $0 \le k \le 2n$. We denote by L the differential operator T acting on the domain

(4.22) $$D(L(\omega)) := \{f \in H^n[0,1] : B_i(\omega,f) = 0,\ i = 1,\ldots,k\},$$

where $H^n[0,1]$ is the Sobolev space of all functions f in $C^{n-1}[0,1]$ with $f^{(n-1)}$ absolutely continuous and $f^{(n)}$ in $L_2[0,1]$, with the usual norm

$$|f|_n = \sum_{i=0}^{n-1} ||f^{(i)}||_\infty + ||f^{(n)}||_2.$$

It is easy to see that L is a random operator (with stochastic domain $D(L)$) from $\Omega \times H^n$ (or from $\Omega \times L_2$) into L_2. We define L^* to be the adjoint operator of $L:D(L) \to L_2$ (with respect to the standard inner product in L_2) with prescribed domain

(4.23) $$D(L^*(\omega)) := \{f \in H^n[0,1] : B_i^*(\omega,f) = 0,\ i = 1,\ \ldots,\ 2n-k\},$$

where

(4.24) $$B_i^*(\omega,f) := \sum_{j=0}^{n-1} \{a_{ij}(\omega) f^{(j)}(a) + b_{ij}(\omega) f^{(j)}(b)\},\ j=1,\ldots,2n-k,$$

are $2n-k$ linearly independent adjoint boundary operators, where a_{ij} and b_{ij} are measurable. It is known that L^* is the formal adjoint of T on $D(L^*)$. It should be noted that if $k < 2n$ in

(4.22), then L^* would not be uniquely determined without specifying the adjoint boundary conditions associated with (4.23). Clearly L^* and $L^\dagger$ depend on the choice of the boundary conditions, and in what follows L^* and $L^\dagger$ should be understood relative to the fixed boundary operators in (4.22) and (4.23).

The operator L is a densely defined closed linear operator from $L_2[0,1]$ into $L_2[0,1]$ and has closed range in $L_2[0,1]$. Therefore the Moore - Penrose inverse $L^\dagger$ is a bounded linear operator on $L_2[0,1]$ onto $D(L)$ equipped with the L_2-topology. On the other hand, L is a bounded linear operator from $D(L)$ equipped with the H^n-topology onto $R(L)$. Hence $(L|D(L)\cap N(L)^\perp)^{-1} = L^\dagger|R(L)$ is continuous from $R(L)$ under the L_2-topology onto $D(L)\cap N(L)^\perp$ equipped with H^n-topology. We observe that the functional $\phi_t(f) := (L^\dagger f)(t)$ is a bounded linear functional for each fixed t since point-evaluation functionals are continuous on H^n. Hence by the Riesz representation theorem there exists a unique $L_2[0,1]$ - function $G(t,\cdot)$ such that

$$(L^\dagger f)(t) = \phi_t(f) = \int_0^1 G(t,s)f(s)ds$$

for all $f \in L_2$ and all t in $[0,1]$. The function $G(t,s)$ is called the <u>generalized Green's function</u> for L. It turns out that $G(t,\cdot)$ and $G(\cdot,s)$ are continuous on $[0,1]$ and that G is jointly continuous on $\{(t,s) : t \neq s\}$; for these and other properties of the (generalized) Green's function for an nth order linear differential operator, see [14, Chapter XIII] and [48].

We now return to the consideration of the <u>random</u> differential operator $L(\omega)$. Under the assumptions of Theorem 2.7, $L^\dagger$ is a <u>random</u> linear operator from $\Omega \times L_2$ into L_2. It follows from above that for each ω, $L^\dagger(\omega)$ is an integral operator with kernel $G(\omega,t,s)$. Because of the continuity properties of $G(\omega,\cdot,\cdot)$ mentioned earlier, we may invoke Lemma 4.1 to conclude that G is in fact a random kernel, which we will therefore call the <u>random generalized Green's function</u>. We have thus established the following proposition.

<u>Proposition 4.6</u>: Let T be the random formal differential operator defined in (4.20) and let L be the operator T restricted to the domain $D(L(\omega))$ defined in (4.22). Let L^* be the adjoint operator with the prescribed domain (4.23). We assume that $\bigcap_\omega D(L(\omega))$ and

$\bigcap_{\omega} D(L^*(\omega))$ are dense in L_2 and that $L^*(\cdot)L(\cdot)$ is separable from L_2 into L_2. Then the generalized inverse $L^\dagger$ is a random integral operator on $\Omega \times L_2$; its kernel G is a random kernel in the sense of Lemma 4.1.

In particular if $N(L) = \{0\}$, then Proposition 4.6 asserts that the Green's function of L is a random kernel.

We now illustrate the use of random generalized Green's functions in connection with nonhomogeneous <u>random boundary</u>- and <u>eigenvalue problems</u>. Let $f:\Omega \to L_2$ be measurable, and consider the boundary-value problem

$$(4.25) \qquad L(\omega)x = f(\omega) \qquad x \in D(L(\omega))$$

Under the assumptions of Proposition 4.6,

$$(4.26) \qquad x(\omega) = L^\dagger(\omega)f(\omega) = \int_0^1 G(\omega,\cdot,s)f(\omega,s)ds$$

is measurable. If f satisfies the compatability condition of the Fredholm alternative (i.e., $f \in N(L^*)^\perp$), then x is the unique solution of (4.25) of minimal L_2-norm. Otherwise, (4.25) is not solvable; in this case x is the least-squares solution of (4.25) of minimal norm.

Now we treat the nonhomogeneous eigenvalue problem. Let λ be a measurable real-valued function and consider

$$(4.27) \qquad L(\omega)x - \lambda(\omega)x = f(\omega), \qquad x \in D(L(\omega)),$$

where $\lambda(\omega)$ is allowed to be an eigenvalue of $L(\omega)$ for some ω. To establish the existence of random solutions (least-squares solutions) and to construct such solutions, we can use one of the following two approaches:

(i) If L is such that the generalized Green's function G_λ of $L - \lambda I$ can be easily constructed, then we obtain directly the (least-squares) solution of minimal norm of (4.27) in the form

$$x(\omega) = (L(\omega) - \lambda(\omega)I)^\dagger f(\omega) = \int_0^1 G_{\lambda(\omega)}(\omega,\cdot,s)f(\omega,s)ds.$$

(ii) Let L be <u>self-adjoint</u> and let G be the random generalized Green's function of L. We denote by P and Q the orthogonal projectors onto $N(L)$ and $R(L)$ respectively. If x is a least-squares solution of (4.27), then $L^\dagger Lx = \lambda L^\dagger x + L^\dagger Qf$ and therefore

(4.28) $$x - Px = \lambda L^{\dagger}x + L^{\dagger}f.$$

Conversely, if x is a solution of (4.28), then

$$L^{\dagger}(Lx - \lambda x - Qf) = L^{\dagger}(Lx - \lambda Qx - Qf) = 0$$

so

$$Lx - \lambda Qx - Qf \in N(L^{\dagger}) = R(L)^{\perp}.$$

But $Lx - \lambda Qx - Qf$ is also in $R(L)$; hence,

$$L(x - Px) - \lambda(x - Px) = Qf,$$

so $x - Px$ is a least-squares solution of (4.27). Finally, if x is a least-squares solution of (4.27), then $x - Px$ is also a least-squares solution. In this sense the equations (4.27) and (4.28) are equivalent. In particular if we seek the least-squares solution x of minimal norm (equivalently, if we restrict the equations (4.27) and (4.28) to $N(L)^{\perp}$), then x is the solution in $N(L)^{\perp}$ of the integral equation

(4.29) $$x = \lambda(\omega)\int_0^1 G(\omega,\cdot,s)x(s)ds + y(\omega,\cdot)$$

where $$y(\omega,\cdot) := \int_0^1 G(\omega,\cdot,s)f(\omega,s)ds.$$

In all of the preceding discussion we have considered the generalized inverse of the differential operator in the topology of L_2. If $D(L)$ is independent of ω, one might also consider the generalized inverse of the bounded random operator $L:D(L) \to L_2[0,1]$, where $D(L)$ bears the Banach-space structure of H^n. Since $N(L)$ is finite dimensional, it has a topological complement, say M, in the Banach space $D(L)$. Thus, in this setting $L^{\dagger}_{M,R(L)}$ is defined. Measurability results could be obtained from Section II.5. Finally, we could equip $D(L)$ with the Hilbert space structure generated by a Sobolev inner product on H^n. However, neither of these alternative approaches is directly connected with the generalized Green's function described above.

4. Iterative Methods for Random Linear Boundary-Value Problems

The purpose of this short section is to illustrate the applicability of iterative methods to least-squares solutions to random linear boundary-value problems. It suffices here to indicate how such problems can be recast in the framework of Chapter III. We consider the formal random differential operator defined in (4.20) subject to the boundary operators in (4.21), and define the operators L and L^* as in Section 3 (see (4.22)-(4.24)). We introduce on the space $H^n[0,1]$ the graph inner products

$$\langle f,g\rangle_{T(\omega)} = \langle f,g\rangle + \langle T(\omega)f,T(\omega)g\rangle$$

and the corresponding norm $||f||_T = \sqrt{\langle f,f\rangle_T}$; we refer to the topology induced by this norm as the graph topology for H^n. As these graph topologies are different for each $\omega\in\Omega$, we cannot use them to define measurability of H^n-valued functions; thus, measurability and randomness for H^n-valued functions and operators should in the rest of this section be understood with respect to the Sobolev topology for H^n, which is induced by the norm

$$|f|_n = \sum_{i=0}^{n-1} ||f^{(i)}||_\infty + ||f^{(n)}||_2 .$$

Clearly, $L(\omega)$ is a bounded linear operator from $D(L(\omega))$ equipped with its graph topology onto $R(L(\omega))$ with the L_2-topology; hence there exists a (graph) adjoint $L_*:\Omega \times L_2 \to H^n$ with

$$\langle L(\omega)f,g\rangle_{T(\omega)} = \langle f,L_*(\omega)g\rangle_{T(\omega)} \quad \text{for} \quad f\in D(L(\omega)),g\in L_2[0,1].$$

For each $\omega, L_*(\omega):L_2\to D(L(\omega))$. It is not hard to show that $N(L_*) = N(L^*)$, $R(L_*) = D(L)\cap N(L)^\perp \subset R(L^*)$ and $L_*=L^*(LL^*+I)^{-1}$, where LL^* is the $2n^{th}$ order differential operator defined by

$$D(LL^*)=\{f\in H^{2n}[0,1]: B_i^*(f)=B_j(T^*f)=0, i=1,\ldots,2n-k \text{ and } j=1,\ldots,k\},\ LL^*f = TT^*f.$$

Since the adjoint of the bounded linear random operator $L:H^n\to L_2$ is a random operator (see Proposition 1.12), the preceding expression for L_* together with Corollary 1.10 and Lemma 1.3 implies that L_* is a random operator. Moreover, the generalized inverse of L, where L is considered as an operator on H^n with its graph topology into L_2 with its natural topology, is bounded; equivalently, the operator L is bounded below, i.e., there exists a constant $m > 0$ such that

$$||Lf|| \geq m||f||_T \quad \text{for} \quad f \in D(L) \cap N(L)^{\perp}.$$

In view of these properties, it follows immediately that the results of Chapter III on iterative and projection methods apply to the random linear boundary-value problem under consideration. In particular, the adaptation of the method of steepest descent to this setting takes the following form. Consider the functional

$$J(\omega,f) := \frac{1}{2}\,||L(\omega)f - g(\omega)||^2$$

defined on Gr D(L), where each $D(L(\omega))$ is equipped with its graph norm.

Then $f(\omega) \in D(L(\omega))$ is a least-squares solution of

$$L(\omega)f(\omega) = g(\omega)$$

if and only if $f(\omega)$ is a solution of the normal equation

$$L_*(\omega)L(\omega)f = L_*(\omega)g(\omega).$$

The method of steepest descent for minimizing $J(\omega,f)$ is defined by the sequence

$$f_{n+1}(\omega) = f_n(\omega) - \alpha_n(\omega)r_n(\omega), \quad n=0,1,2,\ldots$$

where the initial approximation f_o: $\Omega \to H^n$ is a given random variable such that for all $\omega \in \Omega, f_o(\omega) \in D(L(\omega))$,

$$r_n := L_*Lf_n - L_*g_o \in D(L) \cap N(L)^{\perp}.$$

Note that $f_n(\omega) \in D(L(\omega))$ for all n and ω. The step size α_n is chosen to minimize $J(\omega,f_{n+1}(\omega))$; i.e., α_n is given by

$$\alpha_n(\omega) = \frac{||r_n(\omega)||^2}{||L(\omega)r_n(\omega)||^2}.$$

It follows from the randomness of L_* that all f_n are measurable. Theorem 3.1 and Remark 3.2 apply verbatim to our setting.

Various iterative methods that are known to converge to a least-squares solution of random linear equations involving a bounded operator apply to this setting. The projection schemes of Section III.2 can also be adapted in a routine way to the solution of random boundary-value problems.

5. Approximations to Least-Squares Solutions of Random Integral Equations of the First and Second Kind

So far we have described three approaches that can be applied to obtain approximations to least-squares solutions of random integral equations of the second kind and/or the computation of a generalized inverse of $I-\lambda(\omega)K(\omega)$, namely:

(i) The use of the Hurwitz pseudoresolvent kernel; in this procedure, which can be viewed as a perturbation method to obtain an invertible operator, it is necessary to find orthonormal bases for $N(I-\lambda(\omega)K(\omega))$ and $N(I-\overline{\lambda}(\omega)K^*(\omega))$.

(ii) Iterative methods such as the method of steepest descent. Various iterative methods for (least-squares) solutions of linear integral equations of the first and second kind have been studied in [36]; the random analogues of the results of [36] can be obtained in a routine way using tools developed in this paper. A drawback of these methods in the case of second kind equations is the slow convergence if the pseudocondition number of $I-\lambda K$ is large. For first kind equations, iterative methods can be only used in practice in conjunction with regularization or filteration procedures for coping with numerical instability of ill-posed problems.

(iii) Projection methods, using the setting and results of Section III.2.

We now describe a method based on <u>quadrature approximations</u> for the operator $K(\omega,x)$ in (4.1). The convergence properties of quadrature approximations hinge on the interplay between the generalized inverses of the operator $A(\omega,x):=x-K(\omega,x)$ on the spaces $C[0,1]$ and $L_2[0,1]$ (see Lemma 4.2) and the theory of collectively compact operators. These convergence properties have been developed for the deterministic setting in [51]. All the needed tools for the corresponding theory for the random case have been developed in this paper (see especially Section IV.1).

Let the integral operator K in (4.1) be approximated (in the space C) by a quadrature rule with nodes t_{nj} and weights w_{nj}:

$$K_n(\omega,x)(s):=\sum_{j=1}^{n} w_{nj}(\omega)\; k(\omega,s,t_{nj})\; x(t_{nj})$$

and suppose that on the space C

$$K(\omega,\cdot) \text{ is compact and } ||K_n(\omega,x)-K(\omega,x)||\to 0.$$

Then Theorem 4.1 of [51] can be immediately adapted to the random situation. It should be noted that $(I-K_n)^\dagger$ need not converge to $(I-K)^\dagger$; however, a variant $(I-K_n)^\phi$ of $(I-K_n)^\dagger$ converges to $(I-K)^\dagger$ and certain asymptotic relations hold.

For the case when $I-K$ is invertible, the measurability results of this paper give immediately measurability and convergence results for the method of quadrature approximation for random integral equations of the second kind; computer studies of this method have recently been performed by A. T. Bharucha-Reid and M. Christiansen.

6. Applications to Nonlinear Problems

Measurability results for generalized inverses of random operators can be combined with methods of random fixed point theory to prove existence of measurable solutions and to provide stochastic approximation schemes for nonlinear random operator equation containing a noninvertible linear part, either in additive form ("nonlinear alternative problems") or in multiplicative form (e.g., Hammerstein equations $KFx=x$, where $N(K) \neq \{0\}$).

A. Nonlinear Alternative Problems

In several papers (e.g., [37],[38]) Kannan and Salehi study the existence of measurable solutions of "nonlinear alternative problems", i.e., equations of the form

$$Lx = Fx \tag{4.30}$$

where L is a bounded linear random operator with closed range and nontrivial kernel and F a nonlinear random operator which fulfills monotonicity and continuity conditions. To prove the existence of measurable solutions, they use the well-known fact that (4.30) is equivalent to the system of "alternative equations"

$$x = Px + L^\dagger QFx \tag{4.31a}$$

$$(I-Q)Fx = 0 \tag{4.31b}$$

where P is a projector onto $N(L)$ and Q a projector onto $R(L)$. (4.31a) is usually called the "auxiliary equation," and (4.31b), the "bifurcation equation."

To use (4.31) for establishing the existence of measurable solutions, one has first to make sure that all operators involved are random operators.

Kannan and Salehi consider the following setting: L and F map X^* into H, where X is a Banach space with separable dual space X^*, H is a separable Hilbert space and (X,H,X^*) are "in normal position" (i.e., $X^* \subseteq H \subseteq X$ algebraically and topologically with some additional conditions). By invoking measurability results for $L^\dagger$ also for unbounded L or in Banach spaces (cf. Section II.5), we are able to treat alternative problems also in the case where L and F act between two separable Banach spaces X and Y without further conditions on X and Y or where L is a closed operator on a Hilbert space. We give two typical results. Since our objective here is to establish measurability, we assume that the deterministic alternative problem has a solution. Various conditions when this is the case are discussed in Cesari [11] and Hale [26] where numerous additional references can be found.

Theorem 4.7: Let X and Y be separable Banach spaces, $L:\Omega\times X\to Y$ a bounded linear random operator with closed range, $F:\Omega\times X\to Y$ a continuous (not necessarily linear) random operator. Let $M:\Omega\to 2^X$ and $S:\Omega\to 2^Y$ be measurable such that for all $\omega\in\Omega$, $M(\omega)$ and $S(\omega)$ are subspaces of X and Y, respectively, with $X = N(L(\omega)) \oplus M(\omega)$ and $Y = R(L(\omega)) \oplus S(\omega)$.

We assume that for each fixed $\omega_o\in\Omega$, the deterministic problem $L(\omega_o)x = F(\omega_o)x$ is solvable. Then there exists a measurable function $x:\Omega\to X$ such that for all $\omega\in\Omega$, $L(\omega)x(\omega) = F(\omega)x(\omega)$.

Proof: Let $P:\Omega\times X\to X$ be the projector onto $N(L)$ parallel to M, $Q:\Omega\times Y\to Y$ the projector onto $R(L)$ parallel to S. Because of Corollary 2.17, P and Q are random operators. From Theorem 2.16 we conclude that $L^\dagger_{M,S}$ is a random bounded operator. Together with Lemma 1.3 this implies that $P + L^\dagger QF$ and $(I-Q)F$ are continuous random operators.

For each $\omega\in\Omega$, let $H(\omega) := \{x\in X: x=P(\omega)x+L^\dagger(\omega)Q(\omega)F(\omega)x\}$ and $B(\omega) := \{x\in X: F(\omega)x=Q(\omega)F(\omega)x\}$. It follows from the assumptions that $H(\omega)$ and $B(\omega)$ are nonvoid and closed. By results from stochastic fixed point theory (cf. Theorem 6 and its proof in [20]), H and B are measurable, and so is $H\cap B$. Because of our assumption about the solvability of the deterministic alternative problem, each $H(\omega)\cap B(\omega)$ is nonvoid. The existence of a measurable function $x:\Omega\to X$ with $Lx = Fx$ follows now from Theorem 1.8 and the observation that each element of $H(\omega)\cap B(\omega)$ solves $L(\omega)x = F(\omega)x$. □

Remark 4.8: By applying other results about measurability of $L^\dagger$

we could get other versions of Theorem 4.7 under assumptions on L like those in Theorems 2.14 and 2.15. All these results can be easily generalized to the case where F is not defined on all of X but on a stochastic domain (since the corresponding results are available in stochastic fixed point theory, cf. [16],[18],[19],[20]) and where F has weaker continuity properties (e.g., demicontiuity). It should be noted, however, that since L and F are assumed to be continuous, the statement of Theorem 4.7 could also be obtained (without appealing to the measurability of $L^{\dagger}$) from results in stochastic fixed point theory ([19],[65]). This is not the case, however, if we permit L to be unbounded as in the following result, the proof of which is along the same lines as the proof of Theorem 4.7 using Theorem 2.7 instead of Theorem 2.16.

Theorem 4.9: Let X and Y be separable Hilbert spaces, L a closed linear random operator from $\Omega\times X$ into Y with closed range such that the following conditions are satisfied:

(i) $\bigcap_{\omega} D(L(\omega))$ is dense in X;

(ii) $\bigcap_{\omega} D(L^*(\omega))$ is dense in Y;

(iii) $T^*(\cdot)T(\cdot)$ is separable.

Furthermore let $F:\Omega\times X\to Y$ be a continuous random operator. We assume that for each fixed $\omega_o\in\Omega$, the deterministic problem $L(\omega_o)x = F(\omega_o)x$ is solvable.

Then there exists a measurable function $x:\Omega\to X$ such that for all $\omega\in\Omega$, $L(\omega)x(\omega) = F(\omega,x(\omega))$.

B. Iterative Methods for Random Nonlinear Equations

Let X and Y be real or complex Hilbert spaces and let $F:\Omega\times X\to Y$ be a strongly monotone random operator; i.e., there exists a measurable function m with $m(\omega)> 0$ for all ω, such that

$$\mathrm{Re}\langle F(\omega,x_1)-F(\omega,x_2),x_1-x_2\rangle \geq m(\omega)||x_1-x_2||^2. \tag{4.32}$$

We also assume that F is lipschitzian; more precisely, for some measurable function $M:\Omega\to\mathbb{R}$,

$$||F(\omega,x_1)-F(\omega,x_2)|| \leq M(\omega)||x_1-x_2||. \tag{4.33}$$

Let $\alpha:\Omega\to\mathbb{R}^+$ be measurable and define $T_\alpha := I-\alpha F$. T_α is a random operator and

$$\begin{aligned}||T_\alpha x_1-T_\alpha x_2||^2 &= ||x_1-x_2-\alpha(Fx_1-Fx_2)||^2\\ &= ||x_1-x_2||^2-2\alpha\mathrm{Re}\langle x_1-x_2,Fx_1-Fx_2\rangle+\alpha^2||Fx_1-Fx_2||^2.\end{aligned}$$

Thus using (4.32) and (4.33) it follows that

$$||T_\alpha(\omega,x_1)-T_\alpha(\omega,x_2)||^2 \le \{1-2m(\omega)\alpha(\omega)+M^2(\omega)\alpha^2(\omega)\}||x_1-x_2||^2 ,$$

and so for each ω, $T_\alpha(\omega,\cdot)$ is a contraction mapping for any α which satisfies for all ω the condition

$$0 < \alpha(\omega) < \frac{2m(\omega)}{M(\omega)} . \tag{4.34}$$

In this case the sequence

$$x_{n+1}(\omega) = x_n(\omega) - \alpha(\omega)F(\omega,x_n(\omega)), \tag{4.35}$$

where the initial approximation x_o is a measurable function, converges to the unique fixed point x^* of T_α (equivalently the unique solution of $F(x) = 0$). By Lemma 1.3 the sequence $\{x_n\}$ defined by (4.35) is a sequence of random variables; hence x^* is a measurable function. A particular choice of the function α as a measurable function is $\alpha(\omega) = \frac{m(\omega)}{M(\omega)}$.

Conditions (4.32) and (4.33) can be weakened in various ways (see [59]). The parameter $\alpha(\omega)$ can be replaced by a linear operator $B(\omega)$ satisfying suitable conditions. In particular if F has a continuous Fréchet derivative, then $B(\omega)$ can be chosen as a bounded outer inverse for $F'(\omega,x_o(\omega))$. Conditions under which this method converges in the deterministic setting are given in [60]. Random analogues can be derived by taking $B(\omega)$ to be a measurable outer inverse which is possible in view of Theorem 2.25.

REFERENCES

[1] M. Altman, Contractors and Contractor Directions: Theory and Applications, Marcel Dekker, New York, 1977.

[2] G. F. Andrus and T. Nishiura, Fixed points of random set-valued maps, Nonlinear Analysis (to appear).

[3] G. A. Bécus, Successive approximation of solutions of a class of random equations, preprint, University of Cincinnati, 1978.

[4] R. Bellman, Stochastic Processes in Mathematical Physics and Engineering, Proceedings of Symposia in Applied Mathematics, Vol. 16, Amer. Math. Soc., Providence, R.I., 1964.

[5] A. Ben-Israel and T. N. E. Greville, Generalized Inverses: Theory and Applications, Wiley-Interscience, New York, 1974.

[6] F. J. Beutler and W. L. Root, The operator pseudoinverse in control and systems identification, pp. 397-494 in [53].

[7] A. T. Bharucha-Reid, *Random Integral Equations*, Academic Press, New York, 1972.

[8] A. T. Bharucha-Reid, Fixed point theorems in probabilistic analysis, *Bull. Amer. Math. Soc.* 82 (1976), 641-657.

[9] N. Bourbaki, *Espaces Vectoriels Topologiques*, Éléments de Mathématique, Vol. 18, Hermann & Cie., Paris, 1955.

[10] R. C. Brown, Generalized Green's functions and generalized inverses for linear differential systems with Stieltjes boundary conditions, *J. Differential Equations* 16 (1974), 335-351.

[11] L. Cesari, Functional analysis, nonlinear differential equations, and the alternative method, *Nonlinear Functional Analysis and Differential Equations* (L. Cesari, R. Kannan and J. D. Schuur, editors), Marcel Dekker, New York, 1976, pp. 1-197.

[12] G. Debreu, Integration of correspondences, *Proc. 5th Berkeley Symp. Math. Stastic. Prob.*, Vol. II, pt. 1, pp. 351-372; University of Cal. Press, Berkeley, 1967.

[13] J. Dixmier, *Les Algèbres d'Opérateurs dans l'Espace Hilbertien* (Algèbres de von Neumann), Gauthier-Villars, Paris, 1957.

[14] N. Dunford and J. T. Schwartz, *Linear Operators*, Parts I and II, Wiley-Interscience, New York, 1958 and 1963.

[15] H. W. Engl, Fixed Point Theorems for Random Operators on Stochastic Domains,Dissertation, Universität Linz (Austria), 1977.

[16] H. W. Engl, Some random fixed point theorems for strict contractions and nonexpansive mappings, *Nonlinear Analysis* 2 (1978), 619-626.

[17] H. W. Engl, Weak convergence of Mann iteration for nonexpansive mappings without convexity assumptions, *Boll. Un. Mat. Ital.* *(5)* 14-A (1977), 471-475.

[18] H. W. Engl, A general stochastic fixed point theorem for continuous random operators on stochastic domains, *J. Math. Anal. Appl.* 66(1978), 220-231.

[19] H. W. Engl, Random fixed point theorems for multivalued mappings, *Pacific J. Math.* 76 (1978), 351-460.

[20] H. W. Engl, Random fixed point theorems, *Nonlinear Equations in Abstract Spaces* (V. Lakshmikantham, editor), Academic Press, New York, 1978, pp. 67-80.

[21] H. W. Engl, Existence of measurable optima in stochastic nonlinear programming and control, *Berichte des Instituts für Mathematik der Universität Linz* (Austria) 116 (1978), and submitted.

[22] H. W. Engl and M. Z. Nashed, Generalized inverses of random linear operators in Banach spaces (to appear).

[23] H. W. Engl and M. Z. Nashed, Stochastic projectional schemes for random linear operator equations of the first and second kinds (to appear).

[24] I. I. Gikhman and A. V. Skorokhod, *Introduction to the Theory of Random Processes*, W. B. Saunders, Philadelphia, 1969.

[25] C. W. Groetsch, *Generalized Inverses of Linear Operators*, Marcel Dekker, New York, 1977.

[26] J. K. Hale, Applications of Alternative Problems, Lecture Notes, Brown University, Providence, R. I., 1971.

[27] O. Hanš, Generalized random variables, Trans. First Prague Conference on Information Theory, Statistical Decision Functions, Random Processes, Publishing House of the Czechoslovak Academy of Sciences, Prague, 1957, pp. 61-103.

[28] O. Hanš, Inverse and adjoint transforms of linear bounded random transforms, Ibid., pp. 127-133.

[29] O. Hanš, Random operator equations, Proc. 4th Berkeley Symposium on Math. Statist. and Prob., Vol. II, Univ. of California Press, Berkeley, California, 1961, pp. 185-202.

[30] E. Hille and R. S. Phillips, Functional Analysis and Semigroups, Colloquium Publications, Vol. 31, Amer. Math. Soc., Providence, R.I., 1957.

[31] C. J. Himmelberg, Measurable relations, Fundamenta Math. 87 (1975), 53-72.

[32] C. J. Himmelberg and F. S. van Vleck, Multifunctions with values in a space of probability measures, J. Math. Anal. Appl. 50 (1975), 108-112.

[33] A. D. Ioffe, Survey on measurable selection theorems: Russian literature supplement, SIAM J. Control and Optimization 16 (1978), 728-732.

[34] S. Itoh, Nonlinear random equations with monotone operators in Banach spaces, Math. Ann. 236 (1978), 133-146.

[35] M. I. Kadets and B. S. Mityagin, Complemented subspaces in Banach spaces, Russian Math. Surveys 28 (1973), 77-95.

[36] W. J. Kammerer and M. Z. Nashed, Iterative methods for best approximate solutions of linear integral equations of the first and second kinds, J. Math. Anal. Appl. 40 (1972), 547-573.

[37] R. Kannan and H. Salehi, Random nonlinear equations and monotonic nonlinearities, J. Math. Anal. Appl. 57 (1977), 234-256.

[38] R. Kannan and H. Salehi, Random solutions of nonlinear differential equations, Boll. Un. Mat. Ital. (4), 12 (1975), 209-213.

[39] R. Kannan and H. Salehi, Mésurabilité du point fixé d'une transformation aléatoire séparable, C. R. Acad. Sci. Paris Ser. A, 281 (1975), 663-664.

[40] J. J. Koliha, Power convergence and pseudoinverses of operators between Banach spaces, J. Math. Anal. Appl. 48 (1974), 446-469.

[41] J. J. Koliha, Convergence of an operator series, Aequationes Math.16 (1977), 31-35.

[42] M. A. Krasnosel'skii, G. M. Vainikko, P. O. Zabreiko, Ja. B. Rutickii and V. Ja.Stecenko, Approximate Solution of Operator Equations, Nauka, Moscow, 1969; English transl., Wolters-Noordhoff, Groningen, 1972.

[43] K. Kuratowski and C. Ryll-Nardzewski, A general theorem on selectors, Bull. Acad. Pol. Sc. (Sér. Math., Astr. et Phys.) 13 (1965), 397-403.

[44] L. J. Lardy, A series representation for the generalized inverse of a closed linear operator, Atti Accad. Naz. Lincei Rend. Cl. Sci. Fis. Mat. Natur. Ser. VIII, 58 (1975),152-157.

[45] M. D. Lax, Method of moments approximate solutions of random linear integral equations, J. Math. Anal. Appl. 58 (1977), 46-55.

[46] M. D. Lax and W. E. Boyce, The method of moments for linear random initial value problems, J. Math. Anal. Appl. 53 (1976), 111-132.

[47] A. C. Lee and W. J. Padgett, On random nonlinear contractions, Math. Systems Theory 11 (1977), 77-84.

[48] J. Locker, The generalized Green's function for an nth order linear differential operator, Trans. Amer. Math. Soc. 28 (1977), 243-268.

[49] G. Mägerl, A unified approach to measurable and continuous selections, Trans. Amer. Math. Soc. 245 (1978), 443-452.

[50] S. F. McCormick and G. H. Rodrigue, A uniform approach to gradient methods for linear operator equations, J. Math. Anal. Appl. 49 (1975), 275-285.

[51] R. H. Moore and M. Z. Nashed, Approximations to generalized inverses of linear operators, SIAM J. Appl. Math. 27 (1974), 1-16.

[52] W. L. Morse, The dishonest method in stream temperature modeling, Water Resources Research 14 (1978), 45-51.

[53] M. Z. Nashed, editor, Generalized Inverses and Applications, Academic Press, New York, 1976.

[54] M. Z. Nashed, Perturbations and approximations for generalized inverses and linear operator equations, in [53], pp. 325-396.

[55] M. Z. Nashed, Aspects of generalized inverses in analysis and regularization, in [53], pp. 193-244.

[56] M. Z. Nashed, Inverse mapping theorem for random operator equations with applications, to appear.

[57] M. Z. Nashed, Steepest descent for singular linear operator equations, SIAM J. Numer. Anal. 7 (1970), 358-363.

[58] M. Z. Nashed, Generalized inverses, normal solvability, and iteration for singular operator equations, Nonlinear Functional Analysis and Applications (L. B. Rall, ed.), Academic Press, New York, 1971, pp. 311-359.

[59] M. Z. Nashed, A decomposition relative to convex sets, Proc. Amer. Math. Soc. 19 (1968), 782-786.

[60] M. Z. Nashed, Generalized inverse mapping theorems and related applications of generalized inverses in nonlinear analysis, Nonlinear Analysis in Abstract Spaces (V. Lakshmikantham, ed.), Academic Press, New York, 1978, pp. 210-245.

[61] M. Z. Nashed and L. B. Rall, Annotated bibliography on generalized inverses, in [53], pp. 771-1041.

[62] M. Z. Nashed and H. Salehi, Measurability of generalized inverses of random linear operators, SIAM J. Appl. Math. 25 (1973), 681-692.

[63] M. Z. Nashed and G. F. Votruba, A unified operator theory of generalized inverses, in [53], pp. 1-109.

[64] J. Nedoma, Note on generalized random variables, Trans. 1st Prague Conf. Inform. Th., Statist. Dec. Fcts., Random Processes, Prague, 1957.

[65] A. Nowak, Random solutions of equations, Trans. 8th Prague Conf. on Inform. Theory, Statist. Decision Functions, Random Processes, Vol. B, Academia, Prague, 1978, pp. 77-82.

[66] W. M. Patterson, 3rd, Iterative Methods for the Solution of a Linear Operator Equation in Hilbert Space - A Survey, Lecture Notes in Mathematics, Vol. 394, Springer-Verlag, Berlin, Heidelberg, New York, 1974.

[67] W. V. Petryshyn, On projectional-solvability and Fredholm alternative for equations involving linear A-proper operators, Arch. Rational Mech. Anal. 30 (1968), 270-284.

[68] F. Riesz and B. Sz.-Nagy, Functional Analysis, Frederick Ungar Publishing Co., New York, 1955.

[69] R. T. Rockafellar, Measurable dependence of convex sets and functions on parameters, J. Math. Anal. Appl. 28 (1969), 4-25.

[70] I. Singer, The Theory of Best Approximation and Functional Analysis, SIAM Regional Conf. Series in Appl. Math., Philadelphia, 1974.

[71] T. T. Soong, Random Differential Equations in Science and Engineering, Academic Press, New York, 1973.

[72] A. E. Taylor, Introduction to Functional Analysis, John Wiley and Sons, New York, 1958.

[73] C. P. Tsokos and W. J. Padgett, Random Integral Equations with Applications to Life Sciences and Engineering, Academic Press, New York, 1974.

[74] D. H. Wagner, Survey of measurable selection theorems, SIAM J. Control and Optimization 15 (1977), 859-903.

[75] K. Yosida, Functional Analysis, 4th Edition, Springer-Verlag, New York, Heidelberg, Berlin, 1974.

Approximate Solution of Random Nonlinear Equations by Random Contractors*

W. J. Padgett

Department of Mathematics, Computer Science, and Statistics
University of South Carolina, Columbia, South Carolina 29208

Abstract
1. Introduction
2. Definitions and Lemmas
3. Iterative Solution of General Random Nonlinear Operator Equations
4. Application to a Random Discrete Fredholm Equation
5. Application to a Random Nonlinear Integral Equation
References

*Research partially supported by the U.S. National Science Foundation under Grant No. MCS78-02915.

ABSTRACT

Let (Ω,A,P) be a complete probability space, let $\omega \in \Omega$, and suppose that X and Y are Banach spaces. The concept of a random contractor of a random nonlinear operator $U(\omega)$: $X \to Y$ and its usefulness in obtaining existence, uniqueness, and approximation of random solutions to random nonlinear operator equations of the form $U(\omega)x(\omega) = u(\omega)$ are discussed. As applications of the random contractor results, the approximate solutions of random nonlinear Volterra integral equations and random nonlinear discrete systems are considered, and in particular the errors of approximation are obtained.

1. INTRODUCTION

Random operator equations arise frequently in the formulation and analysis of problems in the biological, physical, and engineering sciences, and such equations are often nonlinear by nature. Hence, the development of general techniques for solving random nonlinear operator equations is an important part of current research in mathematical modeling. The recent books by Bharucha-Reid[2] and Tsokos and Padgett[8,9] on random integral equations and the text by Soong[7] on random differential equations may be referenced for many of the results and applications of random operator equations which have been developed through the use of "probabilistic functional analysis".

The problem of approximating the solutions of random nonlinear operator equations was considered briefly by Bharucha-Reid[2] and Tsokos and Padgett[9]. Other results concerning approximate solutions of such equations that have been obtained recently include those of Kusano[3] and Lee and Padgett[5,6]. The method given by Lee and Padgett[5] and applied to general random nonlinear integral equations is the concept of a <u>random contractor</u>. The random contractor theory for the solution of random nonlinear operator equations will be described in this presentation, based on the results of Lee and Padgett[5]. In addition, the results for random contractors will be applied to general random discrete Fredholm systems and random nonlinear integral equations in order to obtain existence, uniqueness, and approximation of solutions. In particular, iteration procedures in appropriate spaces of functions will be considered to approximate the solution of such equations, and the specific error of approximation will be obtained for each type of random equation, generalizing previous results of Tsokos and Padgett[8,9].

2. DEFINITIONS AND LEMMAS

In this section some basic definitions, notation, and lemmas will be stated for reference[2,5,9].

Let (Ω,A,P) be a complete probability measure space, and let $(X,\mathcal{B})$ and $(Y,\mathcal{C})$ be two measurable spaces, where X and Y are Banach spaces and $\mathcal{B}$ and $\mathcal{C}$ are σ-algebras of Borel subsets of X and Y, respectively. The definitions of an X-<u>valued random</u>

variable, random operator, and properties of random operators are used as given by Bharucha-Reid[2].

In this paper random operator equations of the form $T(\omega)x = y(\omega)$ will be considered, where $x \in \mathcal{D}(T)$, $y(\omega)$ is a Y-valued random variable, and $T(\omega)$ denotes the random operator with domain $\mathcal{D}(T) \subset X$. Any X-valued random variable $x(\omega)$ which satisfies the condition $P(\{\omega: \ T(\omega)x(\omega) = y(\omega)\}) = 1$ will be said to be a random solution of the random operator equation $T(\omega)x = y(\omega)$.

Let $L(X,Y)$ denote the collection of all bounded linear operators from a Banach space X into a Banach space Y. It is known that $L(X,Y)$ is a Banach space with norm $||\cdot||$ given by

$$||T|| = \sup_{||x||\leq 1} ||Tx||, \ T \in L(X,Y) \text{ and } x \in \mathcal{D}(T).$$

The following lemma is useful for establishing the measurability of solutions of random operator equations[5].

Lemma 2.1 Assume T is an a.s. continuous random operator from $\Omega \times X$ into $L(X,Y)$. Then $y(\omega) = T(\omega; z(\omega))x(\omega)$ is a Y-valued random variable where $x(\omega)$ and $z(\omega)$ are X-valued random variables.

Next some function spaces which are useful in applications will be defined. In particular, the space of all second-order mean-square continuous stochastic processes on $R_+ = [0,\infty)$ and all second-order (discrete parameter) stochastic processes on the positive integers will be considered.

Definition 2.1 Let $y_j(\omega)$, $j = 1,2,\ldots,n$ be second-order real-valued random variables on (Ω,A,P); that is, $y_j(\omega) \in L_2(\Omega,A,P)$ for each j. The collection of all n-component random vectors $\underline{y}'(\omega) = (y_1(\omega),\ldots,y_n(\omega))$ constitutes a linear vector space if all equivalent random vectors are identified. Define the norm of $\underline{y}$ by

$$||\underline{y}||_{L_2^n} = \max_{1\leq j\leq n} ||y_j||_{L_2} = \max_{1\leq j\leq n} \left(\int_\Omega |y_j|^2 dP\right)^{\frac{1}{2}}.$$

The space of all n-component random vectors $\underline{y}$ with second-order components and norm given by $||\cdot||_{L_2^n}$ is a separable Banach space and will be denoted by $L_2^n(\Omega,A,P)$, or simply L_2^n (Soong[7]).

Definition 2.2 Define the space $C(R_+, L_2^n(\Omega,A,P))$ to be the space of all continuous functions from $R_+ = [0,\infty)$ into $L_2^n(\Omega,A,P)$ with the topology of uniform convergence on compacta. It may be noted that $C(R_+, L_2^n(\Omega,A,P))$ is a locally convex space[10] whose topology is defined by a countable family of semi-norms given by

$$||x(t;\omega)||_m = \sup_{t\in[0,m]} ||x(t;\omega)||_{L_2^n}, \qquad m = 1,2,\ldots .$$

It is known that this type of topology is metrizable and the resulting metric space is a separable Fréchet space.

Definition 2.3[8,9] Denote by $\Phi = \Phi(N, L_2(\Omega, A, P))$ the space of all functions $\underline{x}$ from N, the positive integers, into $L_2(\Omega, A, P)$. That is, for each $n = 1,2,\ldots,$ the value of $\underline{x}$ at n is $x_n(\omega) \in L_2(\Omega, A, P)$. The topology of Φ is the topology of uniform convergence on every set $N_m = \{1,2,\ldots,m\}$, $m = 1,2,\ldots$; that is, $\underline{x}_i \to \underline{x}$ as $i \to \infty$ in Φ if and only if $\lim_{i\to\infty} ||x_{i,n}(\omega) - x_n(\omega)||_{L_2} = 0$ uniformly on every set N_m, $m = 1,2,\ldots$.

Note that Φ is also a locally convex space with the topology defined by the following family of semi-norms:

$$||x_n(\omega)||_m = \sup_{0\le n\le m} ||x_n(\omega)||_{L_2}, \quad m = 1,2,\ldots .$$

In addition, Φ is the space of all second order stochastic processes defined on the set of positive integers. That is, the processes are discrete parameter processes. Also, $C(R_+, L_2^n(\Omega, A, P))$ is the space of all mean-square continuous n-vector valued stochastic processes on R_+.

Finally, the concept of admissibility is defined. Let B and D be Banach spaces. The pair (B,D) is said to be a.s. admissible with respect to a random operator $U(\omega)$ if $U(\omega)(B) \subset D$ a.s.

3. ITERATIVE SOLUTION OF GENERAL RANDOM NONLINEAR OPERATOR EQUATIONS

In this section the random contractor theory for the solution of random nonlinear operator equations as given by Lee and Padgett[5] is discussed and the error of approximation for the general iterative random solution is indicated. First, the definition of a random contractor for a random operator is stated. This is a probabilistic generalization of the notion of a contractor for deterministic operators[1].

Definition 3.1 Let X and Y be separable Banach spaces, and let $U(\omega)$: $\Omega \times \mathcal{D}(U) \to Y$ be a nonlinear random operator, where $\mathcal{D}(U) \subset X$ is the domain of $U(\omega)$. Let $\Gamma(x;\cdot)$: $\Omega \times Y \to X$ be a bounded linear random operator associated with $x \in X$. Then $U(\omega)$ has a random contractor $\Gamma(x;\cdot)$ at $x \in \mathcal{D}(U)$ if there exist positive random variables $q(\omega)$, $0 < q(\omega) < 1$ a.s., and $\gamma(\omega)$ such that

$$||U(\omega)[x + \Gamma(x;\omega)y] - U(\omega)x - y|| \le q(\omega)||y|| \text{ a.s.} \tag{3.1}$$

where $y \in Y$ and $||y|| \le \gamma(\omega)$ a.s.

It follows from (3.1) that

$$[1 - q(\omega)]||y|| \le ||U(\omega)[x + \Gamma(x;\omega)y] - U(\omega)x|| \text{ a.s.} \tag{3.2}$$

so that the random contractor $\Gamma(x;\omega)$ is an a.s. one-to-one mapping.

The random solution of the random operator equation

$$U(\omega)x = 0, \tag{3.3}$$

is considered first, where $U(\omega)$: $\Omega \times \mathcal{D}(U) \to Y$ and X and Y are separable Banach spaces. Later, random nonhomogeneous operator equations will be investigated.

Assume that $U(\omega)$ has a random contractor $\Gamma(x;\omega)$ for $x \in S(x_0;r) = \{x \in \mathcal{D}(U): ||x - x_0|| \le r\}$, where $x_0(\omega) = x_0$ a.s. is a given approximate solution to (3.3). Let

$$x_{n+1}(\omega) = x_n(\omega) - \Gamma(x_n(\omega);\omega)[U(\omega)x_n(\omega)], \quad n = 0,1,\ldots, \tag{3.4}$$

be an iterative procedure for the approximate solution of (3.3). Then the following theorem gives sufficient conditions for the convergence of (3.4) to a random solution of (3.3)[5].

Theorem 3.1 Suppose there exist positive random variables $q(\omega)$, $0 < q(\omega) < 1$ a.s., $\gamma(\omega)$, and $B(\omega)$ and a positive number r such that the following conditions hold:

(i) $||\Gamma(x;\omega)|| \le B(\omega)$ a.s. for $x \in S(x_0;r)$;

(ii) $||U(\omega)x_0|| \le \gamma(\omega)$ a.s.;

(iii) $B(\omega)\gamma(\omega)[1 - q(\omega)]^{-1} \le r$ a.s.; and

(iv) $U(\omega)$ is an a.s. closed operator on $S(x_0;r)$.

Then there exists a random solution $x^*(\omega) \in S(x_0;r)$ for equation (3.3) and the sequence of X-valued random variables (3.4) converges to $x^*(\omega)$ a.s., provided that $U(\omega)x(\omega)$ is a Y-valued random variable when $x(\omega)$ is an X-valued random variable.

Proof: Since $x_0(\omega) = x_0$ a.s. is an X-valued random variable and $\Gamma(x;\omega)$ is an a.s. bounded linear random operator for $x \in S(x_0;r)$, by the assumptions and Lemma 2.1, for each n, $x_n(\omega)$ defined by (3.4) is an X-valued random variable and $x_n(\omega) \in S(x_0;r)$ a.s. for $n = 0,1,2,\ldots$.

Now, letting $y(\omega) = -U(\omega)x_n(\omega)$ in (3.4) and substituting into (3.1), we get

$$\begin{aligned} ||U(\omega)x_{n+1}(\omega)|| &\le q(\omega)\, ||U(\omega)x_n(\omega)|| \text{ a.s.} \\ &\le [q(\omega)]^{n+1}\gamma(\omega) \text{ a.s.} \end{aligned} \tag{3.5}$$

Therefore, for $n > m \ge 0$, using assumptions (i) - (iii) and (3.5),

$$\begin{aligned} ||x_n(\omega) - x_m(\omega)|| &\le \sum_{i=m+1}^{n} ||x_i(\omega) - x_{i-1}(\omega)|| \text{ a.s.} \\ &= \sum_{i=m+1}^{n} ||\Gamma(x_{i-1}(\omega);\omega)U(\omega)x_{i-1}(\omega)|| \text{ a.s.} \\ &\le B(\omega) \sum_{i=m+1}^{n} ||U(\omega)x_{i-1}(\omega)|| \text{ a.s.} \\ &\le B(\omega)q^m(\omega)\gamma(\omega)[1 - q(\omega)]^{-1} \text{ a.s.} \\ &\le r\, q^m(\omega) \to 0 \text{ a.s. as } n \to \infty. \end{aligned} \tag{3.6}$$

Hence, $\{x_n(\omega)\}$ is a.s. a Cauchy sequence and converges a.s. to some $x^*(\omega) \in S(x_0;r)$. But since $U(\omega)x_n(\omega) \to 0$ a.s. by (3.5) and $U(\omega)$ is a.s. closed on $S(x_0;r)$, we have $U(\omega)x^*(\omega) = 0$ a.s. □

Inequality (3.6) yields an estimate of error of approximating the random solution $x^*(\omega)$ by the iteration procedure (3.4). Letting $n \to \infty$ in the usual way the error bound is obtained from (3.6) as

$$||x^*(\omega) - x_m(\omega)|| \le B(\omega)q^m(\omega)\gamma(\omega)[1 - q(\omega)]^{-1} \text{ a.s.} \tag{3.7}$$

The random solution will be unique if other conditions are added in Theorem 3.1.

Theorem 3.2 If, in addition to the hypotheses of Theorem 3.1, inequality (3.1) is satisfied for all $y \in Y$ such that $||\Gamma(x;\omega)y|| \le 2r$ a.s. and $\Gamma(x;\omega)$ is onto for $x \in S(x_0;r)$, then there exists a unique random solution, $x^*(\omega) \in S(x_0;r)$, of equation (3.3).

The proof of Theorem 3.2 is given by Lee and Padgett[5].

Theorems 3.1 and 3.2 remain true if the random contractor is a uniform random contractor which is defined next.

Definition 3.2 The random operator $U(\omega)$ is said to have

(i) a bounded random contractor $\Gamma(x;\cdot)$ if there exists a positive random variable $B(\omega)$ such that $||\Gamma(x;\omega)|| \le B(\omega)$ a.s. for all $x \in \mathcal{D}(U)$;

(ii) a uniform random contractor at $x_0 \in \mathcal{D}(U)$ if

$$||U(\omega)[x + \Gamma(x_0;\omega)y] - U(\omega)x - y|| \le q(\omega)||y|| \text{ a.s.}$$

for all $x \in S(x_0;r)$ and $||y|| \le \gamma(\omega)$ a.s., where $\gamma(\omega)$ is a positive random variable.

To prove Theorems 3.1 and 3.2 with a uniform random contractor at x_0, the iteration procedure (3.4) is modified by replacing $\Gamma(x_n(\omega);\omega)$ with $\Gamma(x_0(\omega);\omega)$ at each $n = 1,2,\ldots$.

The random nonhomogeneous operator equation $U(\omega)x = u(\omega)$ is now considered, where $u(\omega)$ is a (nonzero) Y-valued random variable. If the random operator $U(\omega)$ has a random contractor $\Gamma(x;\omega)$, then the operator $T(\omega)$ given by $T(\omega)x = U(\omega)x - u(\omega)$ has the same random contractor. Then the following theorem holds[5].

Theorem 3.3 Suppose that the a.s. closed random nonlinear operator $U(\omega)$ has a bounded random contractor $\Gamma(x;\omega)$ such that inequality (3.1) holds for all $y \in Y$. Then the random operator equation $U(\omega)x = u(\omega)$ has a random solution for an arbitrary Y-valued random variable $u(\omega)$, and the random solution is unique if $\Gamma(x;\omega)$ is onto.

The proof is similar to that of Theorem 3.1 where the following iteration procedure is used:

$$x_{n+1}(\omega) = x_n(\omega) - \Gamma(x_n(\omega);\omega)[U(\omega)x_n(\omega) - u(\omega)],\ n = 0,1,2,\ldots . \tag{3.8}$$

The error bound for the approximation of the unique random solution is again given by inequality (3.7).

Now, consider random operator equations of the form

$$U(\omega)x = x - F(\omega)x = u(\omega), \tag{3.9}$$

where the random nonlinear operator $F(\omega)$: $\Omega \times \mathcal{D}(F) \to X$, $\mathcal{D}(F) \subset X$, is a.s. closed. Condition (3.1) becomes

$$||F(\omega)y - F(\omega)[x + \Gamma(x;\omega)y] - [I(\omega) - \Gamma(x;\omega)]y|| \le q(\omega)||y|| \text{ a.s.} \tag{3.10}$$

for all $y \in X$, where $I(\omega)$ is the a.s. identity operator. If condition (3.10) replaces (3.1) in Theorem 3.3 and $u(\omega)$ is an arbitrary X-valued random variable, then Theorem 3.3 remains true for equation (3.9). If in fact $F(\omega)$ is a random contraction operator, then it is obvious that $I(\omega)$ is a random contractor and a generalization of the random version of Banach's contraction mapping theorem results[2]. Also, the error bound for approximating the random solution $x^*(\omega)$ of (3.9) is again given by (3.7).

Other results concerning the idea of regular random contractors were given by Lee and Padgett[5], but now specific applications of the random contractor theory will be considered.

4. APPLICATION TO A RANDOM DISCRETE FREDHOLM EQUATION

A "discretized" version of the random Fredholm integral equation

$$x(t;\omega) = h(t;\omega) + \int_0^\infty k(t,s;\omega)f(s,x(s;\omega))ds, \quad t \ge 0,$$

may be obtained[6,8,9] by replacing the integral with a sum of the functions evaluated at discrete points $t_1, t_2, \ldots, t_n, \ldots$. In this section the solution of such a random discrete Fredholm equation of the form

$$x_n(\omega) = h_n(\omega) + \sum_{j=1}^{\infty} c_{n,j}(\omega)f_j(x_j(\omega)), \quad n = 1,2,\ldots , \tag{4.1}$$

is considered. These types of equations may arise in the study of discrete systems; see Tsokos and Padgett[9] for example. As a special case, when $c_{n,j}(\omega) = 0$ a.s. for $j > n$, a random discrete Volterra system is obtained.

In this section, the existence of a random solution will be shown and the approximation of the random solution will be studied by applying the random contractor results from Section 3. Specifically, the exact error of approximation of the random solution of (4.1) will be obtained which generalizes that given on page 148 of Tsokos and Padgett[9].

The following assumptions are made concerning the functions in the random system (4.1). The functions $x_n(\omega)$ and $h_n(\omega)$ are functions from the positive integers N into $L_2(\Omega,A,P)$; that is, $\underline{x}, \underline{h} \in \Phi(N,L_2(\Omega,A,P))$. For each value of $x_n(\omega)$, $n = 1,2,\ldots$, $f_n(x_n(\omega))$ is a scalar, and for each n, $f_n(x_n(\omega)) \in L_2(\Omega,A,P)$. For each n and j in N, $c_{n,j}(\omega)$ is assumed to be in $L_\infty(\Omega,A,P)$ so that $c_{n,j}(\omega)f_j(x_j(\omega)) \in L_2(\Omega,A,P)$. Also, for each n, $c_{n,j}(\omega)$ is such that

$$|||c_{n,j}(\omega)||| = ||c_{n,j}(\omega)||_{L_\infty(\Omega,A,P)}$$

and $|||c_{n,j}(\omega)||| \cdot ||x_j(\omega)||_{L_2(\Omega,A,P)}$ are summable over j for every $\underline{x} \in \Phi$. Thus, for each n there must be an integer k_n such that $c_{n,j}(\omega) = 0$ for $j > k_n$.

Consider the random linear operator $T(\omega)$ given by

$$T(\omega)x_n(\omega) = \sum_{j=1}^{\infty} c_{n,j}(\omega)x_j(\omega), \quad n = 1,2,\ldots, \tag{4.2}$$

for $\underline{x} \in \Phi$. It can be shown that $T(\omega)$ is a.s. continuous from Φ into itself.

Theorem 4.1 Suppose the random discrete equation (4.1) satisfies the following conditions:

(i) The Banach spaces B and D are stronger than Φ such that the pair (B,D) is a.s. admissible with respect to the random linear operator $T(\omega)$ given by (4.2);

(ii) $x_n(\omega) \to f_n(x_n(\omega))$ is an operator on $S = \{x_n(\omega) \in D: \ ||x_n(\omega)||_D \le \gamma\}$ with values in B satisfying

$$||f_n(x_n(\omega)) - f_n(y_n(\omega))||_B \le \lambda||x_n(\omega) - y_n(\omega)||_D$$

for $x_n(\omega)$, $y_n(\omega) \in S$ and a positive constant λ; and

(iii) $h_n(\omega) \in D$.

Then there exists a unique random solution $x_n(\omega) \in S$ of (4.1), provided $\lambda K(\omega) < 1$ a.s. and

$$||h_n(\omega)||_D + K(\omega)||f_n(0)||_B \le \gamma[1 - \lambda K(\omega)] \text{ a.s.},$$

where $K(\omega)$ is the norm of $T(\omega)$.

Proof: By the hypotheses and Lemma 5.1.1 of Tsokos and Padgett[9], $T(\omega)$ is a.s. bounded on B. Define the random nonlinear operator $U(\omega)$ from S into D by

$$U(\omega)x_n(\omega) = h_n(\omega) + \sum_{j=1}^{\infty} c_{n,j}(\omega)f_j(x_j(\omega)), \quad n = 1,2,\ldots . \tag{4.3}$$

To show that $U(\omega)(S) \subset S$ a.s., let $x_n(\omega) \in S$. Then

$$\begin{aligned} ||U(\omega)x_n(\omega)||_D &\le ||h_n(\omega)||_D + K(\omega)||f_n(x_n(\omega))||_B \text{ a.s.} \\ &\le ||h_n(\omega)||_D + K(\omega)||f_n(0)||_B + \lambda K(\omega)||x_n(\omega)||_D \text{ a.s.} \\ &\le \gamma \quad \text{a.s.} \end{aligned}$$

Now, the a.s. identity operator $I(\omega)$ is a random contractor for $U(\omega)$, since for $x_n(\omega)$, $y_n(\omega) \in S$,

$$\begin{aligned} &||U(\omega)[x_n(\omega) + I(\omega)y_n(\omega)] - U(\omega)x_n(\omega) - y_n(\omega)||_D \\ &= ||T(\omega)f_n(x_n(\omega)) - T(\omega)f_n(x_n(\omega) + y_n(\omega))||_D \quad \text{a.s.} \\ &\le K(\omega)\ \lambda||y_n(\omega)||_D \quad \text{a.s.} \end{aligned}$$

where $q(\omega) = \lambda K(\omega) < 1$ a.s. Moreover, $I(\omega)$ is onto on S. Therefore, by Theorem 3.2 there exists a unique random solution $x_n^*(\omega) \in S$. □

Now, consider the error of approximation of the random solution to (4.1) under the hypotheses of Theorem 4.1. Denote $\underline{x}(\omega) = (x_1(\omega), x_2(\omega), \ldots) \in \Phi$. Define the random nonlinear operator $F(\omega)$ on the Banach space D of Theorem 4.1 by

$$F(\omega)\underline{x}(\omega) = \sum_{j=1}^{\infty} c_{n,j}(\omega)f_j(x_j(\omega)).$$

Let $U(\omega)\underline{x}(\omega) = \underline{h}(\omega) + F(\omega)\underline{x}(\omega)$ on S as defined by (4.3), and define $W(\omega)\underline{x}(\omega) = \underline{x}(\omega) - U(\omega)\underline{x}(\omega)$. Then $I(\omega)$ is a (bounded by $B(\omega) = 1$ a.s.) random contractor for the random nonlinear operator $W(\omega)$. Also, taking $\underline{x}_0(\omega) = 0$ a.s. as the initial approximate solution, it can be shown that under the conditions of Theorem 4.1,

$$||W(\omega)\underline{x}_0(\omega)||_D \le \gamma[1 - \lambda K(\omega)] \text{ a.s.}$$

Hence, the error of approximating the random solution $\underline{x}^*(\omega) \in S$ for (4.1) by the iteration procedure

$$\underline{x}^{(m+1)}(\omega) = \underline{x}^{(m)}(\omega) - W(\omega)\underline{x}^{(m)}(\omega), \quad m = 0,1,2,\ldots,$$

is given by

$$\begin{aligned} ||\underline{x}^*(\omega) - \underline{x}^{(m)}(\omega)||_D &\le [\lambda K(\omega)]^m[\gamma(1-\lambda K(\omega))][1-\lambda K(\omega)]^{-1} \text{ a.s.} \\ &= [\lambda\ K(\omega)]^m, \quad m = 0,1,2,\ldots\ . \end{aligned}$$

This generalizes the result on page 148 of Tsokos and Padgett[9] which was obtained only for spaces contained in Φ which are of "bounded variation".

5. APPLICATION TO A RANDOM NONLINEAR INTEGRAL EQUATION

In this section the random contractor theory is applied to obtain the existence and approximation of a solution of a random nonlinear integral equation of the form[4,5]

$$x(t;\omega) = h(t;\omega) + \int_0^t k(t,s,x(s;\omega);\omega)ds, \quad t \ge 0. \tag{5.1}$$

This will be accomplished by simultaneously considering the random linear Volterra integral equation

$$x(t;\omega) = h(t;\omega) + \int_0^t k_1(t,s;\omega)x(s;\omega)ds, \quad t \ge 0. \tag{5.2}$$

The following conditions will be assumed concerning the random kernels $k(t,s,x;\omega)$ and $k_1(t,s;\omega)$ in equations (5.1) and (5.2), respectively: The function $k(t,s,x;\omega)$: $\Delta \times R^n \to L_2^n$, where $\Delta = \{(t,s) \in R_+ \times R_+ :\ 0 \le s \le t < \infty\}$, is continuous and $k(t,s,0;\omega) = 0$ a.s. for every $(t,s) \in \Delta$. The function $k_1(t,s;\omega)$: $R_+ \times R_+ \to L_\infty(\Omega,A,P)$ is continuous in t uniformly in $s \in R_+$ from R_+ into $L_\infty(\Omega,A,P)$, and for each $t \in R_+$, $k_1(t,s;\omega)$ is continuous in s from R_+ into $L_\infty(\Omega,A,P)$. Thus, for each $(t,s) \in R_+ \times R_+$, $k_1(t,s;\omega)\ x(s;\omega) \in L_2^n(\Omega,A,P)$.

Define the random integral operators $T_0(\omega)$ and $T(\omega)$ on $C(R_+,L_2^n(\Omega,A,P))$ by

$$[T_0(\omega)x](t;\omega) = \int_0^t k(t,s,x(s;\omega);\omega)ds \tag{5.3}$$

and

$$[T(\omega)x](t;\omega) = \int_0^t k_1(t,s;\omega)x(s;\omega)ds, \tag{5.4}$$

where the integrals are Bochner integrals. From the conditions on $k(t,s,x(s;\omega);\omega)$ and $k_1(t,s;\omega)$, the Bochner integrals are well defined. Moreover, for each $t \in R_+$,

$[T_0(\omega)x](t;\omega) \in L_2^n(\Omega,A,P)$, $[T(\omega)x](t;\omega) \in L_2^n(\Omega,A,P)$, and $T(\omega)$ is an a.s. continuous linear operator from $C(R_+, L_2^n(\Omega,A,P))$ into itself.

Now using the random contractor principle, the following theorem can be obtained.

Theorem 5.1 Assume that the equations (5.1) and (5.2) satisfy the following conditions:

(i) B is a Banach space contained in $C(R_+,L_2^n(\Omega,A,P))$ stronger than $C(R_+, L_2^n(\Omega,A,P))$ such that (B,B) is admissible a.s. with respect to both $T(\omega)$ and $T_0(\omega)$, where $T(\omega)$, $T_0(\omega)$ are the random integral operators defined by (5.3) and (5.4), respectively;

(ii) $k_1(t,s;\omega)$ and $k(t,s,x(s;\omega);\omega)$ behave as described earlier, and

$$||[T(\omega)(x_1 - x_2)](t;\omega) - [(T_0(\omega)x_1)(t;\omega) - (T_0(\omega)x_2)(t;\omega)]||_B \leq \delta||x_1(t;\omega) - x_2(t;\omega)||_B$$

for every $x_1(t;\omega)$, $x_2(t;\omega) \in Q_\rho = \{x(t;\omega) \in B: \ ||x(t;\omega)||_B \leq \rho\}$, where δ is a positive constant with $0 < \delta < 1$; and

(iii) $h(t;\omega) \in B$.

Then there exists a unique random solution, $x^*(t;\omega) \in Q_\rho$, of the equation (5.1), provided that $\delta + K(\omega) < 1$ a.s. and

$$||h(t;\omega)||_B \leq \rho[1 - \delta - K(\omega)] \text{ a.s.}$$

where $K(\omega)$ is the norm of the random operator $T(\omega)$.

The proof is given by Lee and Padgett[5]. It involves showing that $\Gamma(\omega) = [I(\omega) - T(\omega)]^{-1}$ is a uniform random contractor for the random operator $U(\omega)$ given by

$$[U(\omega)x](t;\omega) = x(t;\omega) - h(t;\omega) - [T_0(\omega)x](t;\omega), \ x(t;\omega) \in Q_\rho. \qquad (5.5)$$

Now, taking $x_0(t;\omega) = 0$ a.s. for all $t \geq 0$ as the initial approximation to the unique random solution of (5.1) and using the iteration procedure (3.4) with the random contractor $\Gamma(\omega) = [I(\omega) - T(\omega)]^{-1}$, the error of approximating the random solution of (5.1) by the mth iterate $x_m(t;\omega)$ can be obtained. Letting $U(\omega)$ be defined by (5.5), by the conditions of Theorem 5.1

$$\begin{aligned} ||[U(\omega)x_0](t;\omega)||_B &= ||[T_0(\omega)0](t;\omega) + h(t;\omega)||_B \\ &\leq ||h(t;\omega)||_B + ||[T_0(\omega)0](t;\omega)||_B \\ &= ||h(t;\omega)||_B \\ &\leq \rho[1 - \delta - K(\omega)] \text{ a.s.} \end{aligned}$$

Also, from the proof of Theorem 5.1, $\Gamma(\omega)$ is a bounded random contractor with $B(\omega) = 1/[1-K(\omega)]$ a.s. Next, since $q(\omega) = \delta/[1-K(\omega)] < 1$ a.s.,

$$\begin{aligned} &B(\omega)[\rho(1 - \delta - K(\omega))][1 - q(\omega)]^{-1} \\ &= \frac{1}{1-K(\omega)} [\rho(1 - \delta - K(\omega))][1 - \frac{\delta}{1-K(\omega)}]^{-1} = \rho \quad \text{a.s.} \end{aligned}$$

Therefore, the error of approximating the random solution $x^*(t;\omega) \in Q_\rho$ of (5.1) by the mth iteration from (3.4) is given by

$$||x^*(t;\omega) - x_m(t;\omega)||_B \leq \rho[\frac{\delta}{1-K(\omega)}]^m \text{ a.s., } m = 0,1,2,\ldots .$$

Hence, more general results on approximating the random solution of random Volterra integral equations than those given on page 71 of Tsokos and Padgett[9] have been obtained in two different respects. First, the equation (5.1) considered here is of more general nonlinear form, and secondly, the iteration technique of Section 3 is utilized.

REFERENCES

1. Altman, M. (1973) Inverse differentiability contractors and equations in Banach spaces, Studia Mathematica 46, 1-15.
2. Bharucha-Reid, A. T. (1972) Random Integral Equations, Academic Press, New York.
3. Kusano, Shinako (1977) Dvoretsky's stochastic approximation in Banach space and random integral equations (abstract), I.M.S. Bulletin 6, 218.
4. Lee, A.C.H. and Padgett, W. J. (1976) On a heavily nonlinear stochastic integral equation, Utilitas Mathematica 9, 123-138.
5. Lee, A.C.H. and Padgett, W. J. (1977) Random contractors and the solution of random nonlinear equations, J. Nonlinear Analysis, Theory, Meth. and Appl. 1, 175-185.
6. Lee, A.C.H. and Padgett, W. J. (1977) Some approximate solutions of random operator equations, Bull. Inst. Math. Academia Sinica 5, 345-358.
7. Soong, T. T. (1973) Random Differential Equations in Science and Engineering, Academic Press, New York.
8. Tsokos, Chris P. and Padgett, W. J. (1971) Random Integral Equations with Applications to Stochastic Systems, Lecture Notes in Mathematics, Vol. 233, Springer-Verlag, Berlin.
9. Tsokos, Chris P. and Padgett, W. J. (1974) Random Integral Equations with Applications to Life Sciences and Engineering, Academic Press, New York.
10. Yosida, K. (1965) Functional Analysis, Springer-Verlag, Berlin.

A Finite Element Method for Random Differential Equations*

Tze-Chien Sun

Department of Mathematics
Wayne State University, Detroit, Michigan 48202

1. Introduction
2. A Finite Element Method
3. Error Estimation
4. Remarks About Possible Extensions of This Method
   Acknowledgments
   References

*Research is partly supported by a Wayne State University faculty research grant and NSF grant MCS76-07200.

1. Introduction

The finite element method has been successfully applied to solving differential equations with various boundary conditions in the past ten years (see, e.g., [6]). Its success is due mostly to its flexibility in the geometry of the boundaries and its simplicity in computation. On the other hand the theory of random differential equations is another branch of applied mathematics with rapid growth in the last decade (see, e.g., [1], [2], [3], [5]). In this paper, we shall bring these two branches together by proposing a finite element method for solving random differential equations. For simplicity of notation and derivation, we shall only apply this method to a one dimensional boundary value problem with random non-homogeneous term. The method is derived in detail in §2. In §3, we shall estimate the error bound of the finite element solution. Remarks about possible extensions to other cases are given in §4.

The purpose of this paper is to demonstrate the applicability of finite element method to random differential equations. An advantage of this method is, as can be seen in §2.V.iv, that we can compute the covariance function very easily from the finite element solution. A more complicated problem is treated with a similar method in a separate paper ([7]).

2. A Finite Element Method

I. Preliminary

Let us first consider the one dimensional Sturm-Liouville problem with a random non-homogeneous term.

$$-(p(x)u'(x,\omega))' + q(x)u(x,\omega) = f(x,\omega) , \quad o \le x \le \ell , \tag{1}$$

with boundary conditions

$$u(o,\omega) \equiv u(\ell,\omega) \equiv o \qquad \text{all } \omega \in \Omega . \tag{2}$$

we shall assume

(i) $p \in C^1[o,\ell]$ and $q \in C[o,\ell]$ are non-random, $p(x) \ge p_{min} > o$ and $q(x) \ge o$ for $o \le x \le \ell$.

(ii) $(\Omega,\mathcal{J},P)$ is a probability space on which a standard Brownian Motion $B(x,\omega)$, $o \le x \le \ell$ is defined and $f(x,\omega) = B'(x,\omega)$ is the standard white noise.

The notation prime always means the derivative with respect to x. For $B'(x,\omega)$ the derivative is in the weak sense, i.e., $\forall g \in C^1[o,\ell]$:

$$\int_o^\ell g(x)B'(x,\omega)dx \overset{\text{def}}{=\!=} g(x)B(x,\omega)\Big|_o^\ell - \int_o^\ell g'(x)B(x,\omega)dx.$$

For any two function u and v of x, we shall use the notation

$$(u,v) = \int_o^\ell u(x)v(x)dx ,$$

and for a random variable $h(\omega)$

$$Eh(\omega) = \text{the expectation of } h = \int_\Omega h(\omega)P(d\omega) .$$

Let $\mathcal{B}$ be the class of Borel subsets of the interval $[o,\ell]$ and $\mathcal{A}$ be the sub-σ-field of $\mathcal{F}$ generated by the Brownian motion process $B(x,\omega)$, and let $\mathcal{M}$ be the class of all real-valued $B \times \mathcal{A}$ measurable functions.

Thus, for $u, v \in \mathcal{M}$, we write

$$E(u,v) = E[\int_o^\ell u(x,\omega)v(x,\omega)dx] = \int_\Omega \int_o^\ell u(x,\omega)v(x,\omega)dxdP$$

if the iterated integral exists.

We shall need the following formula for the white noise

$$E[\int_o^\ell g(x)B'(x,\omega)dx \cdot \int_o^\ell h(y)B'(y,\omega)dy] = \int_o^\ell g(x)h(x)dx \tag{3}$$

for any g, $h \in L^2[o,1]$ (see, e.g., [9]).

Let us define the following Hilbert spaces

$$H = \{ v(x,\omega) \in \mathcal{M} , E[\int_o^\ell v(x,\omega)^2 dx] < \infty \}$$

$$H_B^1 = \{ v \in \mathcal{M} , v(o,\omega) \equiv v(\ell,\omega) \equiv 0 , \quad \text{all } \omega \in \Omega \quad \text{and}$$
$$E[\int_o^\ell ((v^2 + (v')^2)] < \infty \}$$

$$H_B^2 = \{ v \in \mathcal{M} , v(o,\omega) \equiv v(\ell,\omega) \equiv o \quad \text{all } \omega \in \Omega \quad \text{and}$$
$$E[\int_o^\ell (v^2 + (v')^2 + (v'')^2] < \infty \} .$$

It is clear that $H_B^2 \subset H_B^1 \subset H$, and the operator

$$L u = -(pu')' + qu$$

is a linear operator from H_B^2 to H.

II. The Equivalent Variational Problem

Define the bilinear functional on $H_B^2 \times H_B^2$

$$a(u,v) = E(L\,u,v) \quad .$$

It follows from the boundary conditions

$$a(u,v) = E\left[\int_o^\ell (pu'v' + quv) \right]. \tag{4}$$

Since only the first derivatives of u and v appear on the right side, we can extend the definition of $a(u,v)$ to $H_B^1 \times H_B^1$.

For $v \in H_B^1$, define the quadratic functional

$$I(v) = a(v,v) - 2\,E(f,v)$$

$$= E\left[\int_o^\ell (p(v')^2 + qv^2 - 2fv) \right]. \tag{5}$$

The following equivalence theorem can be found (except for the random factor) in most books on variational methods (see, e.g., [4], [6]) .

<u>Theorem 1</u>. Let $u \in H_B^1$.

$$I(u) = \min_{v \in H_B^1} I(v) \tag{6}$$

if and only if $Lu = f$ in the weak sense; i.e.,

$$a(u,v) = E(f,v) \quad \text{all} \quad v \in H_B^1 . \tag{7}$$

Moreover, u exists and is unique.

[Proof]

Suppose u minimizes $I(v)$ for $v \in H_B^1$. Then for any $\epsilon > o$, and for any $v \in H_B^1$

$$I(u + \epsilon v) \geq I(u), \tag{8}$$

which implies

$$-2\epsilon[\, a(u,v) - E(f,v)] + \epsilon^2 a(v,v) \geq o \; . \tag{9}$$

It follows $a(u,v) - E(f,v)$ must vanish, or

$$a(u,v) = E(f,v) \quad \text{all} \quad v \in H_B^1 \quad .$$

Conversely if (7) holds, u must be a stationary point of $I(v)$. However, it follows from (4) and the assumptions on p and q , there exists a constant

$c > o$ such that (see, e.g., p. 42[6])

$$a(v,v) \geq c\, E[\int_o^{\ell} (v^2 + (v')^2)] , \quad \text{for each } v \in H_B^1 .$$

Thus for each $o \neq v \in H_B^1$, $a(v,v) \geq c\, E[\int_o^{\ell} (v^2 + (v')^2)] > o$. This, together with (7), (8) and (9), shows that

$$I(u + \epsilon v) > I(u) \quad \text{all } \epsilon > o , \text{ all } v \in H_B^1 .$$

Hence u is the unique element which minimizes $I(v)$ over H_B^1 .

The existence of u follows from the inequality, for $v \in H_B^1$,

$$\begin{aligned} |E(f,v)| &= |E \int_o^{\ell} B'(x,\omega)v(x,\omega)dx| \\ &= |E \int_o^{\ell} B(x,\omega)v'(x,\omega)dx| \\ &\leq [E \int_o^{\ell} B^2(x,\omega)dx \cdot E \int_o^{\ell} (v'(x,\omega))^2 dx]^{1/2} \\ &\leq \frac{\ell}{\sqrt{2}} [E \int_o^{\ell} (v')^2]^{1/2} \\ &\leq \frac{\ell}{\sqrt{2}} \frac{1}{\sqrt{p_{min}}} [E \int_o^{\ell} (p(v')^2 + qv^2)]^{1/2} \\ &\leq K\, a(v,v)^{1/2} . \end{aligned}$$

It says that $E(f,v)$ is a continuous linear functional in Hilbert space Ha with norm $\|v\|_a = a(v,v)^{1/2}$. It is also easy to show that the norms in Ha and H_B^1 are equivalent, so the two spaces contain the same elements. Then there exists an element $u_o \in Ha$ such that

$$E(f,v) = a(u_o,v) \quad \text{all } v \in Ha .$$

We thus have, for $v \in H_B^1$,

$$\begin{aligned} I(v) &= a(v,v) - 2E(f,v) \\ &= a(v,v) - 2a(u_o,v) \\ &= a(v-u_o,v-u_o) - a(u_o,u_o) . \end{aligned}$$

Therefore $I(v)$ has a minimum when $v = u_o$. Q.E.D.

III. Bases

Subdivide the interval $[o,\ell]$ into N equal subintervals with length

$h = \frac{\ell}{N}$. Define

$$\begin{aligned}\varphi_k(x) &= o & o \le x \le (k-1)h \\ &= [x - (k-1)h]/h & (k-1)h \le x \le kh \\ &= [kh - x]/h + 1 & kh \le x \le (k+1)h \\ &= o & (k+1)h \le x \le \ell \end{aligned} \tag{10}$$

for $k = 1,2, \dots N-1$, and define

$$\Delta B_j(\omega) = B(jh,\omega) - B((j-1)h, \omega) \tag{11}$$

for $j = 1,2, \dots , N$ where $B(x,\omega)$, $o \le x \le \ell$ is the standard Brownian motion. Denote

$$\Psi_{j,k}(x,\omega) = \Delta B_j(\omega) \cdot \varphi_k(x) \tag{12}$$

for $j = 1,2, \dots , N$ and $k = 1,2, \dots , N-1$. It is easy to check each $\Psi_{j,k} \in H_B^1$. Let

$$S^h = \{v^h : v^h = \sum_{j=1}^{N} \sum_{k=1}^{N-1} a_{j,k}\Psi_{j,k} \text{ where } a_{j,k} \text{ are real numbers}\}.$$

Then (i) $\{\Psi_{j,k}\}$ form a basis for S^h ,

(ii) $S^h \subset H_B^1$,

(iii) Dimension of $S^h = N(N-1)$.

For each

$$v^h(x,\omega) = \sum_{j=1}^{N} \sum_{k=1}^{N-1} a_{j,k}\Delta B_j(\omega)\varphi_k(x) \in S^h ,$$

we shall compute

$$I(v^h) = E[\int_o^\ell (p(v^{h\prime})^2 + q(v^h)^2 - 2fv^h)] .$$

(i)

$$E[\int q(v^h)^2]$$

$$= \sum_j \sum_k \sum_p \sum_m a_{j,k}\, a_{p,m} E(\Delta B_j \Delta B_p) \int q(x)\varphi_k(x)\varphi_m(x)dx$$

$$= \sum_{j=1}^{N} \sum_{k=1}^{N-1} (c_{k,k-1}a_{j,k}a_{j,k-1} + c_{k,k}a_{j,k}a_{j,k} + c_{k,k+1}a_{j,k}a_{j,k+1})h$$

where $c_{k,k+i} = \int_o^\ell q(x)\varphi_k(x)\varphi_{k+i}(x)dx$, $i = -1,0,1$, with the understanding that $c_{k,k+i} = o$ if either k or $k+i$ is less than 1 or bigger than N .

Here we have used the fact

$$E(\Delta B_j \Delta B_p) = \delta_{j,p} h \ .$$

Note that $c_{k,k+i} = 0(h)$ and it can be integrated easily if $q(x)$ is not too complicated . Also note that $c_{k,k+i} = c_{k+i,k}$.

(ii)

$$E[\int p(v^{h'})^2]$$

$$= \sum_j \sum_k \sum_p \sum_m a_{j,k} a_{p,m} E(\Delta B_j \Delta B_p) \int p(x)\varphi_k'(x)\varphi_m'(x)dx$$

$$= \sum_{j=1}^{N} \sum_{k=1}^{N-1} (d_{k,k-1}a_{j,k}a_{j,k-1} + d_{k,k}a_{j,k}a_{j,k} + d_{k,k+1}a_{j,k}a_{j,k+1})j$$

where $d_{k,k+i} = \int p(x)\varphi_k'(x)\varphi_{k+i}'(x)ds$, $i = -1,0,1$, with the understanding that $d_{k,k+i} = o$ if k or $k+i$ is less than one or bigger than N . Note $d_{k,k+i} = 0(1/h)$. Also $d_{k,k+i} = d_{k+i,k}$.

(iii)

$$E[\int fv^h]$$

$$= \sum_j \sum_k a_{j,k} E[\Delta B_j \int_o^{\ell} \varphi_k(x)B'(x,\omega)dx]$$

$$= \sum_j \sum_k a_{j,k} E[\Delta B_j (-\int_o^{\ell} \varphi_k'(x)B(x,\omega)dx)]$$

$$= \sum_j \sum_k a_{j,k} E[\Delta B_j (-\int_{(k-1)h}^{kh} B(x,\omega)dx + \int_{kh}^{(k+1)h} B(x,\omega)dx)]\frac{1}{h}$$

$$= \frac{h}{2} \sum_{j=1}^{N} (a_{j,j} + a_{j,j-1}) ,$$

where $a_{N,N} = o$ and $a_{1,o} = o$.

Combining (i), (ii) and (iii), we have

$$I(v^h) = \sum_j \sum_k (c_{k,k-1}a_{j,k}a_{j,k-1} + c_{k,k}a_{j,k}a_{j,k} + c_{k,k+1}a_{j,k}a_{j,k+1})h$$

$$+ \sum_j \sum_k (d_{k,k-1}a_{j,k}a_{j,k-1} + d_{k,k}a_{j,k}a_{j,k} + d_{k,k+1}a_{j,k}a_{j,k+1})h$$

$$- \sum_j (a_{j,j} + a_{j,j-1})h \quad .$$

IV. Stiffness Matrix

To find $u^h = \sum_{j=1}^{N} \sum_{k=1}^{N-1} \bar{a}_{j,k} \Delta B_j \varphi_k \in S^h$ such that $I(u^h) = \min_{v \in S^h} I(v)$

we set

$$\frac{\partial I(v^h)}{\partial \bar{a}_{j,k}} = o \quad \text{for } j = 1,2, \ldots , N \text{ and } k = 1,2, \ldots , N-1 ,$$

we have $N(N-1)$ linear equations for $N(N-1)$ unknowns $\bar{a}_{j,k}$, $j = 1,2, \ldots , N$ and $k = 1,2, \ldots , N-1$. By a straightforward computation, we have the following matrix equation

$$K\, a = F \tag{13}$$

where

(i) $a^T = (\bar{a}_{1,1} , \bar{a}_{1,2} , \ldots , \bar{a}_{1,N-1} , \bar{a}_{2,1} , \bar{a}_{2,2} , \ldots , \bar{a}_{2,N-1} , \ldots , \bar{a}_{N,1}, \bar{a}_{N,2} , \ldots , \bar{a}_{N,N-1})$.

(ii)

$$K = \begin{bmatrix} D_1 & & & 0 \\ & D_2 & & \\ & & \ddots & \\ 0 & & & D_N \end{bmatrix}$$

with

$$D_j = D = \begin{bmatrix} A_1 & B_1 & & & 0 \\ B_1 & A_2 & B_2 & & \\ & B_2 & \ddots & \ddots & \\ & & \ddots & \ddots & B_{N-2} \\ 0 & & & B_{N-2} & A_{N-1} \end{bmatrix} , \quad j = 1,2, \ldots N ,$$

$$A_k = 2(c_{k,k} + d_{k,k}) \qquad k = 1,2, \ldots , N-1$$

$$B_k = (c_{k,k+1} + c_{k+1,k}) + (d_{k,k+1} + d_{k+1,k}), \qquad k = 1,2, \ldots , N-2 ,$$

(iii) $F^T = (E_1^T, E_2^T, \ldots, E_N^T)$

with

$E_1^T = (1, 0, \ldots, 0)$ (N - 1)-vector

$E_j^T = (0, \ldots, 0, 1, 1, 0, \ldots, 0)$ (N - 1)-vector with ones in the $(j-1)^{th}$ and j^{th} places, $j = 2,3, \ldots, N-1$,

$E_N^T = (0, \ldots, 0, 1)$ (N - 1)-vector .

V. Remarks About This Method

(i) Note that the matrix K is diagonal in the submatrices D_k's and all the D_k's are the same and are tri-diagonal. The equation $Ka = F$ is really N systems of $N - 1$ equations

$$D_j a_j = E_j \qquad j = 1,2, \ldots, N ,$$

where $a_j^T = (\bar{a}_{j,1}, \bar{a}_{j,2}, \ldots, \bar{a}_{j,N-1})$. This is due to the fact that the white noises are independent over disjoint intervals. Therefore the computation procedure is quite simple.

(ii) The method also works if the non-homogeneous term is replaced by $f(x,\omega) = r(x)B'(x,\omega)$ where, say, $r(x) \in C^1[o,\ell]$. Only now the integrals in $E[\int fv^h]$ is more complicated.

(iii) It might seem that we could use the finite element method on x only, for each fixed ω . This does not work for the case in remark (ii), for we would then have to compute integrals like $\int g(x)B(x,\omega)dx$ for some function g. In general, we can not compute this integral explicitly. We have to approximate this integral anyway, so we may as well use the finite element method on x and ω simultaneously.

(iv) From

$$u^h(x) = \sum_{j=1}^{N} \sum_{k=1}^{N-1} \bar{a}_{j,k} \Delta B_j(\omega)\varphi_k(x) ,$$

we can easily compute the value of $Cov(u^h(x) , u^h(y))$ for any $o < x, y < \ell$.

$$\begin{aligned}
&\mathrm{Cov}(u^h(x),\ u^h(y)) \\
&= \sum_j \sum_k \sum_s \sum_t \bar{a}_{j,k}\, \bar{a}_{s,t}\, \varphi_k(x)\, \varphi_t(y)\, E(\Delta B_j\, \Delta B_s) \\
&= \frac{\ell}{N} \sum_j \sum_k \sum_t \bar{a}_{j,k}\, \bar{a}_{j,t}\, \varphi_k(x)\, \varphi_t(y) \\
&= \frac{\ell}{N}\, \varphi(x)\, A^T A\, \varphi^T(y) \qquad (14)
\end{aligned}$$

where $\varphi(x) = (\varphi_1(x), \varphi_2(x), \ldots, \varphi_{N-1}(x))$

$$A = \begin{pmatrix} \bar{a}_{ij} \end{pmatrix}_{\substack{i = 1, \ldots, N \\ j = 1, \ldots, N-1}}$$

Or, since, from (i), $a_j^T = D^{-1}E_j$, we can write

$$\mathrm{Cov}(u^h(x),\ u^h(y)) = \frac{\ell}{N}\, \varphi(x) D^{-1} E E^T D^{-1} \varphi^T(y) \qquad (15)$$

where
$$E^T = \begin{pmatrix} E_1^T \\ E_2^T \\ \vdots \\ E_N^T \end{pmatrix} .$$

VI. An Example

Consider the simplest case

$-u'' + u = B'(x,\omega)$ with $u(o) \equiv u(1) \equiv o$

Now $p(x) \equiv q(x) \equiv 1$. So

$$\begin{aligned}
c_{k,k+i} &= \int_o^1 \varphi_k(x)\varphi_{k+i}(x)dx = {}^{2h}/_3 && i = o \\
&= {}^{h}/_6 && i = -1,1. \\
d_{k,k+1} &= \int_o^1 \varphi_k'(x)\varphi_{k+i}'(x)dx = {}^{2}/_h && i = o \\
&= -{}^{1}/_h && i = -1,1.
\end{aligned}$$

Hence

$$A_k = 2(c_{k,k} + d_{k,k}) = {}^{4h}/_3 + {}^{4}/_h \qquad k = 1,2, \ldots , N - 1,$$

and
$$\begin{aligned}
B_k &= 2(c_{k,k+1} + c_{k+1,k}) + 2(d_{k,k+1} + d_{k+1,k}) \\
&= {}^{h}/_3 - {}^{2}/_h \qquad k = 1,2, \ldots , N - 2 .
\end{aligned}$$

Let us solve the system for $N = 4$ and $h = 1/4$. Now

$$A_k = 52/3 \qquad k = 1,2, \ldots , N-1 ,$$

and

$$B_k = -\,95/12 \qquad k = 1,2, \ldots , N-2 .$$

So we have

$$\begin{aligned} \frac{52}{3}\bar{a}_{j,1} - \frac{95}{12}\bar{a}_{j,2} \qquad\qquad &= e_{j,1} \\ -\frac{95}{12}\bar{a}_{j,1} + \frac{52}{3}\bar{a}_{j,2} - \frac{95}{12}\bar{a}_{j,3} &= e_{j,2} \qquad j = 1,2,3,4 \\ -\frac{95}{12}\bar{a}_{j,2} + \frac{52}{3}\bar{a}_{j,3} &= e_{j,3} \end{aligned}$$

where

$$e_{1,1} = 1 , e_{1,2} = o , e_{1,3} = o ,$$

$$e_{2,1} = 1 , e_{2,2} = 1 , e_{2,3} = o ,$$

$$e_{3,1} = o , e_{3,2} = 1 , e_{3,3} = 1 ,$$

$$e_{4,1} = o , e_{4,2} = o , e_{4,3} = 1 .$$

The solutions are

$$\bar{a}_{1,1} = 0.0783, \bar{a}_{1,2} = 0.0452 , \bar{a}_{1,3} = 0.0207 ,$$

$$\bar{a}_{2,1} = 0.1236, \bar{a}_{2,2} = 0.1442 , \bar{a}_{2,3} = 0.0659 ,$$

$$\bar{a}_{3,1} = 0.0659, \bar{a}_{3,2} = 0.1442 , \bar{a}_{3,3} = 0.1236 ,$$

$$\bar{a}_{4,1} = 0.0207, \bar{a}_{4,2} = 0.0452 , \bar{a}_{4,3} = 0.0783 .$$

Hence

$$u^h(x,\omega) = (\bar{a}_{1,1}\Delta B_1 + \bar{a}_{2,1}\Delta B_2 + \bar{a}_{3,1}\Delta B_3 + \bar{a}_{4,1}\Delta B_4)(4x) , \quad 0 \le x \le \frac{1}{4}$$

$$\begin{aligned} u^h(x,\omega) = {} & (\bar{a}_{1,1}\Delta B_1 + \bar{a}_{2,1}\Delta B_2 + \bar{a}_{3,1}\Delta B_3 + \bar{a}_{4,1}\Delta B_4)4(\tfrac{1}{2} - x) \\ & + (\bar{a}_{1,2}\Delta B_1 + \bar{a}_{2,2}\Delta B_2 + \bar{a}_{3,2}\Delta B_3 + \bar{a}_{4,2}\Delta B_4)4(x - \tfrac{1}{4}) , \quad \tfrac{1}{4} \le x \le \tfrac{1}{2} \end{aligned}$$

where

$$\Delta B_j = B(\tfrac{j}{4},\omega) - B(\tfrac{j-1}{4},\omega) \qquad j = 1,2,3,4 .$$

Note $u^h(x,\omega)$ is symmetric with respect to $x = \frac{1}{2}$. The covariance function of $u^h(x)$ can be computed according to (14) or (15) .

3 Error Estimation

I. An Error Bound for the Finite Element Solution u^h

Let $G(x,y)$ be the Green's function of the boundary value problem (1) and (2) (see [8], p. 273). It is easy to check that

$$u(x,\omega) = \int_o^{\ell} G(x,y)f(y,\omega)dy \tag{16}$$

is a weak solution of (1) and (2) in the sense

$$a(u,v) = E(f,v) \quad \text{for all} \quad v \in H_B^1 .$$

Hence it is the solution of (6) and (7) .

It is also known that

$$G(x,y) \in C^2 \tag{17}$$

in the closed triangle $o \le x \le y \le \ell$ and $G(x,y) = G(y,x)$ all x and y ([8], p. 273). Note that although $G(x,y)$ is continuous in $o \le x,y \le \ell$, its derivatives have discontinuities along the diagonal $x = y$.

From (16), it can be shown

$$u'(x) = \int_o^{\ell} \frac{\partial}{\partial x} G(x,y)B'(y,\omega)dy .$$

Let us approximate $\frac{\partial}{\partial x} G(x,y)$ by

$$H(x,y) = \sum_{j=1}^{N} \sum_{k=1}^{N} b_{j,k} I_{\Delta j,k}(x,y)$$

such that

$$b_{j,k} \cdot h^2 = \iint_{\Delta j,k} H(x,y)dxdy = \iint_{\Delta j,k} \frac{\partial}{\partial x} G(x,y)dxdy \tag{18}$$

where $\Delta j,k = [(j - 1)h,jh) \times [(k - 1)h,kh)$ and $I_{\Delta j,k}$ is the characteristic function of $\Delta j,k$. (17) and (18) imply that

$$\left|\frac{\partial}{\partial x} G(x,y) - H(x,y)\right| \le Mh \tag{19}$$

for some $o < M < \infty$, for all (x,y) except $(x,y) \in \Delta j,j$ $j = 1,2, \ldots , N$.

Hence

$$U(x,\omega) = \int_o^{\ell} H(x,y)B'(y,\omega)dx = \sum_j \sum_k b_{j,k} I_{\Delta j}(x) \Delta B_k(\omega)$$

is an approximation of $u'(x)$, where $\Delta j = [(j - 1)h,jh)$.

Let

$$u_o(x) = \int_o^x U(x,\omega)dx \; .$$

Then, for each ω , $u_o(x)$ is piece-wise linear and

$$u_o(o) \equiv u_o(\ell) \equiv o \quad \text{all} \quad \omega \in \Omega \; ,$$

by (18). Hence $u_o(x) \in S^h$. Now

$$E\int_o^\ell [u'(x) - u_o'(x)]^2 dx$$

$$= E\int_o^\ell \left[\int_o^\ell \left(\frac{\partial}{\partial x} G(x,y) - H(x,y)\right) B'(y,\omega)dy\right]^2 dx$$

$$= \int_o^\ell \int_o^\ell \left(\frac{\partial}{\partial x} G(x,y) - H(x,y)\right)^2 dydx \qquad \text{by (3)}$$

$$= \sum_j \sum_k \iint_{\Delta j,k} \left(\frac{\partial}{\partial x} G(x,y) - H(x,y)\right)^2 dydx$$

$$= \sum_{j\neq k} \sum M^2 h^4 + \sum_j 4\bar{M}^2 h^2 \qquad \text{by (19)}$$

$$\leq \ell^2 M^2 h^2 + 4\ell\bar{M}^2 h \leq Ch$$

where $\bar{M} = \sup_{o\leq x,y\leq \ell} \left|\frac{\partial}{\partial x} G(x,y)\right|$ and $C \geq \ell^2 M^2 h + 4\ell\bar{M}^2$.

It, then, implies

$$E\int_o^\ell p(x)(u'(x) - u_o'(x))^2 dx \leq p_{max}Ch$$

and by an inequality of Poincare type

$$E\int_o^\ell q(x)(u(x) - u_o(x))^2 dx \leq q_{max}\ell^2 Ch \; .$$

Thus, we have

$$a(u - u_o \; , \; u - u_o) \leq Kh$$

where $K = p_{max}C + q_{max}\ell^2 C$.

It is easy to show (p. 39, [6])

$$a(u - u^h \; , \; u - u^h) = \min_{v\in S^h} a(u - v \; , \; u - v) \; .$$

Since $u_o \in S^h$, so we have shown

$$a(u - u^h \; , \; u - u^h) \leq Kh \; . \qquad (20)$$

II. The Convergence Rate of $u^h(x)$

For each fixed x , the convergence rate of $u^h(x)$ to $u(x)$ is bounded by

$$
\begin{aligned}
E(u(x,\omega) - u^h(x,\omega))^2 \\
&= E[\int_o^x (u'(s,\omega) - (u^h(s,\omega))')ds]^2 \\
&\leq E[\int_o^x (u' - (u^h)')^2 \cdot \int_o^x 1^2] \\
&\leq \ell\, E\int_o^\ell (u' - (u^h)')^2 \\
&\leq \frac{\ell}{P_{min}} a(u - u^h, u - u^h) \\
&\leq \frac{\ell K}{P_{min}} h \ . \qquad (21)
\end{aligned}
$$

4 Remarks About Possible Extension of This Method

(I) It is easy to see that this method can be extended without much difficulty to similar two-dimensional problems; e.g.,

$\Delta u = f(x,y,\omega)$ in R, a region with a smooth boundary B,

and $u = o$ on B

where Δ is the Laplacian and $f(x,y)$ is the two-dimensional white noise.

(II) Suppose $f(x)$ is a random process which does not have orthogonal increments like the white noise; e.g., f may be a second-order stationary process. In this case we have to find an orthogonal basis for the space of random variables with finite variance generated by linear combinations of the variables $f(x)$'s to replace the ΔB_j's .

(III) It is possible to apply finite element method to problems (1) and (2) when q is also random. Note that the coefficient q appears in the Green's function $G(x,y)$ non-linearly. We may have to find an orthogonal basis ξ_n in the space of random variables with finite variance generated by all non-linear functions of q's, using the homogeneous polynomials introduced by N. Wiener in [9]. We have to enlarge the basis in (12) by including ΔB_j , φ_k and ξ_n's . In this case, the derivation of the matrix equation and error bound estimation are much more complicated. However, the method is still similar to the present one (see [7]).

(IV) It is also interesting to apply this method to problems with other boundary conditions.

Acknowledgements: The author would like to express his thanks to Professors D. Yen and S. William for helpful suggestions, and to Professor Williams for his comments on the final manuscripts.

References

1. L. Arnold, Stochastic Differential Equations, John Wiley (1974).

2. A. T. Bharucha-Reid (editor), Probability Methods in Applied Mathematics, Vol. I, II, III, Academic Press (1973).

3. A. T. Bharucha-Reid, Random Integral Equations, Academic Press (1972).

4. S. G. Mikhlin, The Problem of the Minimum of a Quadratic Functional, Holden-Day (1965).

5. T. T. Soong, Random Differential Equations in Science and Engineering, Academic Press (1973).

6. G. Strang and G. Fix, An Analysis of the Finite Element Methods, Prentice-Hall (1973).

7. T. C. Sun, A finite element method for random differential equations with random coefficients, to appear.

8. V. S. Vladiminov, Equation of Mathematical Physics, Marcel Dekker (1971).

9. N. Wiener, Non-linear Problems in Random Theory, M.I.T. Press (1958).

Author Index

A
Altman, M., 206, 214, 221
Andrus, G. F., 152, 206
Anselone, P. M., 28, 35
Apartsin, A. S., 104
Arnold, L., 15, 24, 224, 237

B
Bakushinskii, A. B., 104
Barry, M. R., 15, 24
Bécus, G. A., 4, 8, 9, 11, 12
Bellman, R., 194, 206
Ben-Israel, A., 167, 206
Beutler, F. J., 169, 206
Bharucha-Reid, A. T., 4, 12, 29, 30, 31, 35, 97, 104, 151, 152, 153, 158, 174, 189, 194, 207, 212, 213, 217, 221, 224, 237
Billingsley, P., 55, 83, 121, 126
Blankenship, G., 62, 67, 83
Bourbaki, N., 179, 207
Boyce, W. E., 15, 21, 24, 128, 140, 148, 209
Brézis, H., 93, 104
Brockett, R. W., 83
Brown, R. C., 207
Bruck, R. E., Jr., 94, 95, 104

C
Cesari, L., 204, 207
Chow, P. L., 38, 39, 40, 42, 43, 46, 47, 48
Chung, K. L., 119, 126
Cogburn, R., 54, 62, 84
Cozzarelli, F. A., 4, 12

D
Dawson, D. A., 38, 48
Debreu, G., 207
DiMasi, G., 109, 114, 118, 121, 122, 126
Distefano, N., 3, 4, 11, 12
Dixmier, J., 179, 180, 207
Dolph, C. L,, 94, 104
Doob, J. L., 56, 84
Driml, M., 82, 84, 96, 98, 102, 104
Duda, R. O., 70, 84
Dunford, N., 197, 207
Dupač, V., 96, 104

E
Engl, H. W., 152, 156, 161, 164, 174, 176, 177, 178, 179, 183, 204, 205, 207

F
Federgruen, A., 117, 126
Fiacco, A. V., 97, 100, 104
Fix, G., 224, 226, 237
Franklin, J. N., 103, 104

G
Geman, S., 52, 53, 65, 84
Gikhman, I. I., 157, 207
Greville, T. N. E., 168, 206
Groetsch, C. W., 169, 207

H
Hale, J. K., 204, 208
Hanš, O., 30, 31, 35, 98, 101, 104, 152, 158, 161, 163, 208
Hart, P. E., 70, 84
Hartman, P., 57, 84
Hersh, R., 54, 84
Hille, E., 33, 35, 208
Himmelberg, C. J., 159, 208

I
Infante, E. F., 67, 84
Ioffe, A. D., 159, 208
Itoh, S., 152, 208

K
Kadets, M. I., 167, 208
Kammerer, W. J., 202, 208
Kannan, R., 29, 31, 35, 97, 99, 102, 103, 104, 105, 152, 204, 208
Karhunen, K., 4, 12
Keller, J. B., 11, 12
Khasminskii, R. Z., 54, 61, 62, 84
Kiefer, J., 82, 84
Kohler, W. E., 15, 21, 24, 54, 62, 84
Koliha, J. J., 175, 176, 208
Krasnosel'skii, M. A., 183, 186, 208
Krasulina, T. P., 96, 105
Kuo, H. H., 42, 48
Kuratowski, K., 159, 208
Kusano, S., 212, 221
Kushner, H. J., 83, 84, 109, 111, 113, 114, 115, 117, 118, 119, 121, 122, 124, 125, 126

L
Lardy, L. J., 169, 208
Lax, M. D., 15, 24, 128, 139, 147, 209
Lee, A. C., 152, 209, 212, 213, 214, 215, 216, 217, 219, 220, 221
Lindvall, T., 121, 127
Ljung, L., 83, 84
Locker, J., 194, 197, 209
Lybrand, E. R., 60, 84

M
McCormick, G. P., 97, 100, 104
McCormick, S. F., 97, 104, 182, 209
Mägerl, G., 159, 209

Mikhlin, S. G., 226, 237
Minty, G. J., 94, 104
Mitropolsky, Iu. A., 50, 84
Mityagin, B. S., 167, 208
Monro, S., 96, 105
Moore, R. H., 190, 202, 203, 209
Morse, W. L., 153, 209

N
Nashed, M. Z., 29, 31, 35, 152, 153, 159, 161, 163, 164, 168, 169, 172, 173, 174, 176, 177, 178, 179, 181, 182, 183, 184, 190, 192, 194, 202, 203, 206, 207, 208, 209
Nedoma, J., 82, 84, 96, 102, 104, 156, 210
Nishiura, T., 152, 206
Nowak, A., 152, 205, 210

P
Padgett, W. J., 12, 20, 25, 151, 152, 209, 210, 212, 213, 214, 215, 216, 217, 218, 219, 220, 221
Papanicolaou, G. C., 54, 62, 67, 83, 84
Patterson, W. M., 210
Petryshyn, W. V., 183, 210
Phillips, R. S., 33, 35, 208

R
Rall, L. B., 153, 194, 209
Richardson, J. M., 11, 12
Riesz, F., 173, 210
Robbins, H., 96, 105
Rockafellar, R. T., 164, 210
Rodrique, G. H., 182, 209
Root, W. L., 169, 206
Rosenblatt, M., 54, 84
Ross, S. M., 120, 126
Rozanov, Yu. A., 55, 84
Rutickii, Ja. B., 183, 186, 208
Ryll-Nardzewski, C., 159, 208

S
Sakrison, D. J., 83, 85, 96, 105
Salehi, H., 29, 31, 35, 97, 152, 155, 159, 161, 163, 169, 172, 173, 183, 204, 208, 209
Schultz, G., 20, 25
Schwartz, J. T., 197, 207
Schweitzer, P. J., 117, 126
Singer, I., 164, 210
Skorohod, A. V., 97, 105, 121, 127, 157, 207
Soong, T. T., 15, 16, 20, 25, 71, 151, 153, 210, 212, 213, 221, 224, 237
Sorour, E. S., 83, 86
Stecenko, V. Ja., 183, 186, 208
Strang, G., 224, 226, 237
Stratonovich, R. L., 62, 85
Strook, D. W., 107, 108, 110, 112, 113, 126
Sun, T. C., 224, 236, 237
Sutherland, P., 99, 102, 105
Syski, R., 16, 25
Sz-Nagy, B., 173, 210

T
Taylor, A. E., 192, 210
Tsokos, C. P., 12, 20, 25, 151, 153, 210, 212, 214, 217, 218, 219, 221

V
Vainberg, M. M., 91, 105
Vainikko, G. M., 183, 186, 208
van Vleck, F. S., 208
Varadhan, S. R. S., 107, 108, 110, 112, 113, 126
Venter, J. H., 98, 99, 105
Vladiminov, V. S., 234, 237
Volkonskii, V. A., 54, 85
Vorobyev, Y. V., 128, 130, 147
Votruba, G. F., 168, 169, 209
Vrkog, I., 60, 85

W
Wagner, D. H., 159, 210
Wassan, M. T., 70, 82, 85
Weiss, R., 28, 35
White, B. S., 62, 85
Wiener, N., 46, 48, 236, 237
Wolfowitz, J., 82, 84

Y
Yosida, K., 210, 213, 221

Z
Zabreiko, P. O., 183, 186, 208

Subject Index

A
Admissibility, 214
Alternative equations, 203
Approximation(s)
bounded state space, 111
dishonest, 11
of fixed points, 28, 30
of least-squares solution, 202
of mean square continuous functions, 21
quadrature, 202
second order, 11
stochastic, 52, 69, 75, 89, 100
successive, 4, 8, 9, 11, 12
Asymptotic stability, 52, 67
of averaged equation, 52, 67
Auxiliary equation, 203
Averaged equation, 52, 60, 70, 82
Averaging, method of, 50, 101

B
Bifurcation equation, 203
Bilinear random equation, 42
Brownian notion, 39

C
Causal Green's function, 3
Collectively compact operators, 28
Completely regular process, 54
Condition number, 103
Contraction operator(s), random, 30
Contractor, random, 214
Control function, 108, 110
Creep function, 3
Critical value, 76

D
Diffusion, steady-state, 2
Diffusion limit, 62
Displacement, transverse, 3
Domain, stochastic, 155
Dynamic programming equation, 110, 114

E
Evolution equation(s), 39
random, 38, 46
Exponentially stable, 68

F
Finite element method, 224
Fixed point(s)
approximation of, 28
random, 30
Fixed point equations, 28
approximate, 28
Fréchet derivative, 31
Functional(s)
Convex, 90
Gateaux differential of, 90
gradient of, 90
subdifferential of, 93
weakly lower semicontinuous, 90

G
Galerkin-Petrov method, 184
Galerkin's method, 40, 41, 128
Generalized inverses of linear operators, 166
Gradient descent procedure, 75, 88, 89, 96
Green's function, 2, 3, 136
random generalized, 197

H
Hammerstein equations, 203
Heat conduction, 2
Heat equation, 38
Hurwitz pseudoresolvent, random analogue of, 192

I
Ill-posed problem, 103
Inner inverse, 169
Isolated solution, 28
random, 32
Isotropic medium, 2
Itô equation(s), 40, 59, 108

K
Kiefer-Wolfowitz procedure, 82
Kolmogorov equation, forward, 19

L
Langevin-Navier-Stokes equation(s), 46
Liouville's equation, 16
Load, transversal, 3

M
Markov control, 117
Measurable selectors, 159
Medium, isotropic, 2
Method of averaging, 50, 101
Method of moments, 128
Method of steepest descent, 182
Moments, method of, 129
Moore-Penrose equations, 166, 168, 178
Moore-Penrose inverse, 167
of random operator, 172

N
Neumann series, 6
Nice stochastic process, 133

O
Operator(s)
collectively compact, 28
contraction, random, 30
monotone, 92

Operator(s) (continued)
random, 12, 29, 97, 155
Ornstein-Uhlenbeck process, 40
Outer inverse(s), 169
measurability of, 178

P
Perturbation method, modified, 46
Polski condition, 186
Projectors, measurability of, 164, 177

R
Random boundary value problems, 23, 135, 198
iterative methods for, 200
Random contraction operator(s), 30
Random contractor(s), 214
bounded, 216
uniform, 216
Random differential equation(s), 51, 52, 56, 59, 60, 61, 62, 64, 66, 68, 70, 89, 97, 101, 108, 224
finite element method for, 224
Random eigenvalue problems, 198
Random element, 29
Random endomorphism, 29
Random evolution equation(s)
linear, approximation of, 38
nonlinear, 46
Random fixed point equation(s), 31
approximate, 31
Random initial value problem(s), 15, 130
Random integral equation(s), 3, 12, 131, 133, 141, 189, 217, 219
approximations to least-squares solutions, 202
Random nonlinear equations
iterative methods for, 205
Random operator(s), 12, 29, 97, 155
adjoints of, 162
measurability of, 163
bounded, 29
continuous, 29
Fréchet derivative of, 31
generalized inverses of, 161
measurability of, 161, 172, 175
linear, 29
measurability of inverses of, 161
separable, 30, 157
Random operator equation(s), 12, 88, 97, 181, 213, 214, 216
iterative methods for, 181
projection methods for, 183
Random solution(s), 30, 97, 213
approximate, 34
Random variable(s)
generalized, 29, 155
X-valued, 29
Regression function, 101
Regularization method, 88, 94, 95
for random equations, 103
Robbins-Monro procedure, 89, 96, 103, 104

S
Skorokhod imbedding, 121
Skorokhod topology, 121
Solution(s)
approximate, 93
generalized, 4
least-squares, 181
random, 30, 97, 214
weak, 4
Steepest descent, method of, 182
Stochastic approximation(s), 52, 69, 75, 89, 100
Averaging approach to, 77
Stochastic domain, 155
Stochastic least-squares method, 186
Stochastic projection scheme, 184
Strong measurability, 179
Strong mixing, 50, 53, 54
Successive approximations, method of, 4, 8, 9, 11, 12
Sturm-Liouville problem, 224
Submartingale problem, 122
Steady-state diffusion, 2

T
Tight measures, 121
Transverse displacement, 3

U
Uniform asymptotic method, 46

W
Weak convergence, 121, 125
Wiener space, abstract, 39, 41, 42, 43
Wiener-Hermite expansion, 46